INTELLIGENCE BEHIND

THE UNIVERSE!

INTELLIGENCE BEHIND

THE UNIVERSE!

plus

TECHNICAL SUPPLEMENT

- QUANTUM GRAVITATION.

by

R. D. Pearson.

The Headquarters Publishing Co.Ltd.
5 Alexandria Road
London, W134 ONP

December 1990

ISBN 0 947823 21 2

PRINTED IN ENGLAND

CONTENTS

PART I

INTELLIGENCE BEHIND THE UNIVERSE!

PART II

TECHNICAL SUPPLEMENT (T.S.)

REFERENCES

Plus 7 illustrations in PART I and 24 in PART II. List P iv

LIST OF FIGURES.

"INTELLIGENCE BEHIND THE UNIVERSE!"

A BRIEF NOTE ABOUT THE CONTENTS

The main topic of this book is written for the **LAY READER,** but a Technical Supplement is provided for those having a scientific bent. Ideas to solve the perplexing and long standing difficulties about gravitation spin-off into an explanation for the "paranormal" as well! Then amazingly, from a base of theoretical physics, out comes support for the basic faiths man has held since antiquity! Without any doubt the puzzling way light and matter behave on a scale of size smaller than the atom point to an underlying intelligence pervading all matter and space. Not only was the universe created; it is being continuously regenerated at every instant. Not only that, several systems of matter can co-exist interpenetrating each other, so giving scientific support to the idea that people have a separate consciousness or soul.

This SECOND EDITION follows one specially prepared to supplement a lecture concerning **QUANTUM GRAVITATION** presented to a packed audience of physicists at Leeds University on 18th January 1990. A lively discussion ensued.

PART II TECHNICAL SUPPLEMENT - QUANTUM GRAVITATION

Because gravitation is closely linked with the main topic a simplified version of the new theory of quantum gravitation is appended as a technical section. This requires a knowledge of school "A" level physics and mathematics for its understanding. PART II should appeal particularly to engineers, physicists, and mathematicians because they all have an understanding of Newtonian mechanics.

A major spin-off provides clues to the possibility of harnessing a vast new totally non-polluting source of energy.

First Short Edition Distributed   20 February 1989

SYNOPSIS

At the dawn of the age of science this infant discipline seemed a threat to the established Church. For centuries their incompatibilities seemed to widen. But the advent of quantum mechanics promises to heal the rift.

A problem of long standing in quantum theory is posed by the curious way the most minute particles of matter behave. All elements, like iron, when sub-divided until they can be divided no further, are found to be made up of identical atoms. Yet these in turn are composed of sub-atomic particles which do not move like objects of large size. If, for example, some small holes exist in a mask enclosing a source of sub-atomic particles, so they can be made to fly out in all directions at random, then most particles will hit the mask as expected. But any which happen to go through one of the holes move in manner which suggests they pass through *all* the other holes as well simultaneously! Such particles seem to move as if spread out like waves on a pond. Yet sub-atomic particles cannot be waves since they also behave as if they are little hard balls when they hit each other. They act as if they are waves whilst moving through empty space, only collapsing into tiny particles when they hit some obstruction like a screen. So sometimes they behave as if they are waves and at other times as particles. This behaviour is known as "wave-particle duality".

All known attempts at resolution of the dilemma this poses contain internal contradictions or demand impossible assumptions. But the physics which is throwing up these unsatisfactory explanations has a purely atheistic base. A new solution is described, however, based on a common-sense logic which is argued to be totally free from such objections. It homes in on a picture which is inconsistent with the atheistic view. Instead it comes out highly supportive of a creationist scenario and shows that the idea that humankind has an indestructible seat of consciousness or "soul" is consistent with the new extension to physics. Yet wave-particle duality is only one of two key factors of a combined theory leading to exciting new horizons. Both are therefore dealt with in considerable detail.

The other unresolved problem is posed by gravitation. The established view is that Einstein's theory called "General Relativity" is the best in existence. Unlike quantum theory, which concerns the very small range of size, Einstein's relativity theories concern the very large scale.

Unfortunately his theories are not internally consistent and even devoted adherents say they are inconsistent with quantum theory. For over sixty years physicists have been struggling to reconcile the two but despite every twist and turn they have made no solution to the dilemma has emerged. Yet toward the end of his life Einstein himself was suggesting that he could have been on the wrong track after all!

An amazingly simple solution is presented based on an extension to the ideas of Newton. He said that the gravitational force between two massive objects was proportional to the mass of one multiplied by that of the other as well as being affected by their distance of separation. In the new theory energy replaces mass, but this energy has added to it the energy due to any speed of motion. Also space, based on the concepts of quantum theory, is full of invisible particles so that it behaves rather like air. A gas, such as air, consists of many molecules which are like little balls flying about all over the place and bumping into one another. They cannot be seen individually because they are so tiny, but the overall effect is that a gas can flow freely as a fluid. And just as air is compressed by gravity toward the surface of a planet, so that it is thinner at high altitude, so is space. A difference is that space flows without any frictional drag. So space is treated as a "compressible superfluid", an idea which does not seem to have appeared in any prior literature concerning gravitation. These two ideas in combination make all the difference, so that a successful theory of quantum gravitation is seen to emerge. The compressibility of space has the same effect as the "curved-spacetime" of Einstein's theory of gravitation for which he is so famous. But the new theory matches the experimental observations just as well. At the same time all the contradictions and difficulties associated with Einstein's theory vanish like the morning mist!

Then it transpires that other universes of invisible matter could interpenetrate our own without resistance. All need to be supported on an intelligent computer-like background which permeates all space, yet permits space and matter to flow through it as if they had an independent existence. All universes now appear as deliberately contrived illusions but only ours provides gravitation. The others borrow ours so their planets are anchored to our own. Then as a consequence they appear to have different laws of physics. The end results are predictions consistent with the so-called "paranormal", psychic phenomena and spirituality.

DEDICATION

I feel I must dedicate this book to three people. My parents set the stage by providing an environment in my early years which favoured creativity. My father was a physics schoolteacher of an unusual kind. He built all his own apparatus in his home workshop, using it at school to teach by experiment. He ran classes after school for his boys, showing them how to build and fly model aeroplanes. His hobby workshop was an exciting place for a boy, both in early childhood as well as in his teens. Such a background provides a wonderful start for a career in science or engineering.

I also wish to put on record the back up I have had from my wife Margaret for which I am most grateful.

ACKNOWLEDGEMENTS.

I am greatly indebted to Sam Nicholls, BSc.(Phys), a newly graduated physicist of the University of Leeds. He said, "Thank goodness somebody has come up with an explanation for gravitation we can all understand." Then he organised the lecture delivered at that University on 18.1.90. This was the first time the extended Newtonian approach was presented. It is one of the two major concepts this book is based upon.

I wish also to express my gratitude to three eminent professional engineers who have tried to help by obtaining a value judgement of the new theory of gravitation via contacts in the discipline of physics. (They have not seen the other major concept dealing with wave-particle duality). They are:-

Dr. G.B.R. Feilden, CBE, FEng, FRS
Dr. J.H. Horlock, ScD, FEng, FRS (V.C. Open University)
Dr. R.P. Shreeve (Director, Turbopropulsion, Monterey)

All the physicists contacted were firm believers in the establishment view and were highly critical. However, all criticisms made were shown to be invalid and in no case was this response negated.

I wish also to place on record my thanks for help in

presentation by my step-son, lecturer in business management, Mr. Roger Baty, a mathematician, Mr. P. Bentley, Mr. R.F. Taylor, editor of "Two Worlds" and
Mr. R.W. Austin, BSc ARCS, another professional engineer.

FURTHER POSITIVE SUPPORT FROM A PHYSICIST

In addition I wish to place on record my thanks to Professor J.P. Vigier (Paris). In a letter on behalf of "Physics Letters A" he gave support to a section of the theory concerning "gravitational potential energy". It is:-

(Reply to my submission dated July 16th 1987)

Aug. 2nd 1987

The last version of your Letter now seems to me to be correct as far as our referees and myself can see. However, as I told you, the problems discussed therein are not within the range of subjects discussed in Physics Letters A and are more suitable for GRG or some journal like *Fundamenta Scientiae*. I feel that supporters of the Big Bang theory (of which I am not) should discuss the energy problem at creation time (if there is one) and your contribution should not be ignored. (You can utilise this statement if you want and be published). I hope you will succeed in this.

Yours sincerely,   J. Vigier.

BACKGROUND OF THE AUTHOR.
and Origin of this Book.

The author, born in March 1925, began his career as a scientific officer at the National Gas Turbine Establishment, Whetstone Leicester, where he was involved with the emerging technology of jet engines. But at the time his real love was for an invention he originated as a student in the period 1941 to 1946. He ultimately called it a "Gas Wave Turbine" because it used intense pressure waves.

He left to restart its development as a private venture and after three years of single-handed effort made one to work. Then as a result of support given by Dr. G.B.R. Feilden FRS, a full size prototype was built. This surprised everybody by starting at the very first attempt to whisk up to full speed. Unfortunately the supporting firm, Ruston and Hornsby Ltd., ran into financial difficulties so the project had to be abandoned. A paper presented at Rolls Royce premises, Derby, won the Institution of Mechanical Engineers graduates prize for the year, but this is all the effort achieved.

He transferred to a firm in Gloucester where he carried out research on turbochargers for large marine diesel engines in collaboration with the Universities of Manchester and Liverpool, led by Professor J.H. Horlock.

As a result he then transferred to academic life about 1962 under Professor Horlock, head of the School of Engineering Science at Liverpool. Then a few years later he transferred to the University of Bath. Subsequently he was invited by the now Vice Chancellor J.H. Horlock of the Open University, to write a chapter about the engine and associated pressure exchangers for the Memorial volumes being prepared in honour of the late Professor Benson.
{ Editors Horlock and Winterbone(209) }

In 1977 he attempted to launch an expedition to try out the idea of a new kind of marine energy farm. This is touched upon in this book because it combines with proposed applications of the new theory to help solve the developing global problem. The author was assisted by E. Marsden in the attempt to generate enthusiasm for raising the necessary funds. It took the form of displays at two separate international "Energy Show" exhibitions. The required funding did not materialise, however, although a minor Canadian project resulted which aimed to assist in waterweed control of inland lakes and waterways.

Then in 1984 the author's first invention came up again unexpectedly. Others had re-invented the same idea but had not succeeded in making an engine which worked. A "Wave Rotor Workshop" was organised by Dr. R.P. Shreeve at Monterey, for bringing expertise together in an attempt to resolve the difficulties. The way the design problems had been solved in the old engine were described in two papers presented at this conference. As a result, a keen interest was expressed by Dr. G. Walters, head of new engine projects at Pratt and Whitney. Then other approaches came from Canada. A Canadian funded project emerged, assisted by the expertise of Pratt and Whitney and the link-up was finalised on the very day the space-shuttle "Challenger" blew up. It happened during the flight from Vancouver to Florida. Was this an ill omen?

Subsequently the author was invited to emigrate to Canada to restart development. Nine million Canadian dollars were promised from a new "Tax Credit Scheme" to finance design and development.

Six months after resigning his post in 1985 and before actually setting foot in Canada, disaster struck. The entire scheme was cancelled since it had in general attracted the wrong kind of applicant, with the result that 4,000 million dollars of Canadian taxpayer's money had to be written off! So forced early retirement allowed time for other things.

The author had already become interested in physics as a result of publications about the "Big Bang" theory of creation. Some aspects looked wrong. The still unpublished article to which Professor Vigier referred, in the letter quoted in "Acknowledgements", was a critique of Tryon(123). In 1984 Professor Tryon described how physicists were invoking "negative gravitational potential energy" to balance the positive mass-energy of all matter in the universe so that everything could arise from nothing. Then he discussed the "negative pressure of the vacuum". Both arguments lead to incorrect predictions, as will be shown.

They were also concepts totally at variance with the author's understanding. He had the advantage of a different background. This was based on Newtonian physics, thermodynamics and fluid mechanics. So he was able to look at the problem from a new angle, by extending Newtonian physics. These were the triggers which led to the study summarised in this book. As will be shown, a simple solution emerged, easy to understand and visualise in comparison with established physics. The reader can judge its success by its freedom from the numerous internal contradictions which still bedevil the established approach.

The connection with spirituality arose as a natural consequence of the new theory. A new explanation of the phenomenon of "wave-particle duality" needed to be incorporated. The combination threw new light on what up to now has been termed the "paranormal", bringing it within the scope of physics. As a result the author abandoned his previous atheistic convictions.

This book is the end result!

PART 1

1 INTRODUCTION

1.1 SETTING UP THE PROBLEM

Why are we here? Has life any purpose or did it all
just happen by accident?   Are all the so-called
"paranormal" happenings all so much hot air, or have
they a true scientific explanation?   Is there any
substance to the idea that humans have some form of
conscious existence after physical death? These are all
questions everybody asks. We shall explore them by
extending scientific methodology to look for plausible
answers. Clearly science does not yet have all the
answers.   It does not yet offer a satisfactory
explanation for gravitation, nor are the explanations
for the strange phenomenon of wave-particle duality in
any way satisfactory. In the case of paranormal or
psychic phenomena no scientific theory appears to exist.
Indeed the problem has hardly even been addressed in
scientific terms. Could it be that things have been
missed which could unify the whole, both that which is
sensed and unsensed, to provide a simultaneous
explanation for everything? We shall be attempting to
find answers to all these questions by treating them as
a single holistic problem.
   These are not just academic questions, the answers
to which would merely satisfy our curiosity. Correct
answers could lead to valuable spin-off. In the past
new scientific insights have invariably led to new
developments.   These have materially improved the

standards of life. Sadly, they have also increased the
destructive potential of wars. All things seem to have
negative aspects running counter to their positive
values.

In the present age the crying need is for new
insights which could help solve the mounting global
problem. Owing to the growth of population, depletion
of resources and pollution, the projections which have
been made point toward a global disaster scenario. No
matter how man struggles and turns to ward it off, at
the moment a major population crash occurring about the
middle of the next century seems the inevitable
consequence of present trends. The hope must be
therefore to find new insights from new science which
can help avert this tragedy. Compared with this need,
the exploitation of new discovery to further improve
living standards seems trivial and inappropriate. If
anything of value turns up during our search for
answers, then some space will be given to evaluating its
potential. But let us return to the main theme, setting
up the problem we hope to solve.

If the Universe was an unplanned accident and life-
forms similarly just happened, then there is no ultimate
purpose in anything. On this basis some people have
said that the life we experience is all there is. We
just have to make the most of it whilst it lasts, then
we black out into eternal oblivion. On this basis it
matters little if the world's life-carrying capacity is
ultimately destroyed. The belief that nothing can exist
beyond the range of our senses therefore tends to blunt
the drive to think in a long-term manner. If therefore
this view is held by the majority and happens to be
wrong, it could have strong negative consequences. It
is important to know, therefore, whether this view is
true or false, if only for practical reasons.

Countering this view, however, are the faiths
to which all children are exposed. The existence of a
Creator of the universe is postulated and all life is
claimed to be part of a divine plan. Most religions
also assert the existence of a spiritual side to man.

A spirit form living on after death is postulated, or at least some ultimate resurrection followed by eternal survival.

Running counter to this again are the apparent findings of the physical sciences. The established view is in support of an accidental origin. Paul Davies(104) in his *The Accidental Universe* explains this view very clearly. The universe just happened to be created in a gigantic explosion called the "Big Bang" due to a huge "quantum fluctuation" arising in empty space. As a result all the energy of which the universe consists arose from pure nothingness. In so-called "classical physics", which deals with matter on the directly observable scale, the intangible substance called "energy" is assumed to be indestructible. As Einstein proved, the "mass" from which objects are made is equivalent to a certain, and extremely large amount, of energy. Modern quantum theory deals with the same kind of energy but in microscopic detail.

We are not talking here about the so-called "energies" or "force-fields" existing around ancient shrines. Nor are we talking about the positive or negative energies which are said to attract or repel people from one another. These can be ascribed to "psychic energies" of an abstract nature. It is not suggested that psychic energies do not exist, indeed later it will be suggested that they perform an essential role in structuring matter. The energy of classical physics, however, is of a totally different nature. It can be defined as "physical energy" and is the building substance of the universe responsible for both matter and the motion of matter. Where the term "energy" is subsequently used without the adjective "psychic", it is the physical kind to which reference is being made. It is the variety which needs to be supplied, for example, to lift a weight. All forms of physical energy can in principle be interchanged. They can all be expressed in the same units, and the idea of energy lends itself to incorporation in mathematical theories. For example, the number of units of, say, electrical energy needed to boil a kettle of water can be accurately calculated. Yet at a deeper level nobody is ever likely to really know what energy is. At this level it is as abstract as the psychic kind.

Physical energy must not be confused with the so-called energies of the psychic type. It is the latter which are claimed to emanate from hidden watercourses, oil and mineral deposits. These seem detectable because dowsers can exploit them in quite miraculous ways. They have proved time and again to be able to locate the substances for which they are searching. Indeed Uri Geller is so talented in this art that he has become a multi-millionaire by using it to locate crude oil deposits. But this cannot be a real or physical form of energy. It is a psychic energy form. As our quest develops we shall home in on an explanation which shows why psychic energy cannot be measured. We are going to show that it is an abstract kind of energy having no units. It cannot be measured like physical energy and so has to be considered as something entirely different. We shall be taking a close look at this strange psychic energy to uncover exactly what it is and how it works.

There is plentiful evidence that the physical sciences are only addressing a part of nature. It is reasonable to suggest, therefore, that an unsensed part may exist which might be connected with the alternative approach. For example, many people admit to experiencing strange happenings which seem to conflict with all the laws of nature science has uncovered. I have myself experienced such things. Since first-hand experience is the most convincing readers may be prepared to bear with me as I recount a few of these.

My first occurred when I was 21 years old, about a fortnight before sitting for my final degree examinations. I had studied at the local Chesterfield Technical College part-time whilst working through an apprenticeship. Simultaneously I had been developing my first invention. It was a new type of engine, a kind of substitute for a gas turbine, whilst the latter was still in its infancy, the year being 1946. The idea had gradually developed with my increasing technical know-how and then, using my father's well- equipped modelling workshop, I had just completed the first somewhat crude experimental machine. Then I had to leave it alone with some reluctance. The testing would have to wait until after the exams were over as I dare not let it spoil my chances. Then one day, as I sat in the college library studying, I suddenly experienced an irresistible urge to go to the town reference library about a mile walk away. I had not been near it for years and had no need to go there. But I just had to go, even though time was short and I needed to study.

On reaching the entrance hall, I noticed an interesting looking journal whose name I had not seen before. It was called "The Oil Engine and Gas Turbine". I opened it in the middle without turning any pages. Immediately a title flashed into my brain, "Pressure Exchanger for a Locomotive Gas Turbine". It was an article written on behalf of the Brown Boveri Co., Switzerland. I thought, "What a good name "Pressure Exchanger" would have made for my engine." As I read the text my face grew longer and longer. The article seemed to be a good description of my own invention. It was even more advanced than mine in some respects.

This was information I needed to be made aware of, despite the disappointment it entailed. What was it that homed me in on this, even to causing me to open the magazine at exactly the right page? All coincidence? Most people would put it down to this but I know I experienced a feeling I could not resist and no physical explanation can account for that.

The same thing happened twice in 1979. I was working with a friend, Eric Marsden, to prepare for a stand won at the second "Energy Show". Ours was entitled "ENERGY FARMS IN OCEANIC GYRES". The idea was to try and obtain support for an expedition for an experiment in the area of the Sargasso Sea. This is the centre of a rotating body of ocean and possesses natural confining properties for floating seaweed. The idea was to cultivate floating seaweed as an energy crop. Some could be just dried and the rest converted to a substitute crude oil on site. Cultivation only required the artificial upwelling of nutrient-rich seawater from the deeps and a careful analysis had shown the energy balances to be very favourable. It would be highly economic because for one thing the ocean itself provided confinement, so avoiding a high cost. Important advantages were a fish harvest obtained as a result of plankton absorbing about half the nutrient supply and a beneficial effect on the global greenhouse problem. Most of the plankton was sunk as excreta as a result of predation in the food web leading to harvestable fish. This caused carbon dioxide to be absorbed from the atmosphere and ultimately trapped on the seabed. It worked out that for every ton of carbon dioxide released by combustion of product oil between 1.4 and 1.8 tons would be absorbed from the atmosphere. So with this system the more energy used by burning product oil, the more the global greenhouse problem would be eased. No other solution could offer such advantages and yet be economically viable, so we thought we stood a good chance of making the impact needed to attract funds. We were very enthusiastic as we thought we were on to something of considerable importance.

We were working flat out to meet the deadline and a large amount of information had to be located and processed so that meaningful predictions could be calculated. On several occasions I was completely stuck, being unable to find what I required. Then suddenly I felt I had to leave everything and visit the library. I went in, walking round shelves at random. Then in an area quite unsuited to my search, or so I thought, there was the information at the very page of a book selected and opened at random! This happened twice during three weeks.

I have had other experiences but will recount only one more. I was alone and eating my evening meal when I suddenly remembered needing to make a telephone call. On return I found my spectacles sitting near my plate right in front of me. At the same instant I had a feeling that my mother (deceased) was saying, "Look what a mess the table is in." It seemed funny because I remembered putting the glasses away in their case only half an hour earlier. So I thought my memory was playing tricks. On opening the case, however, I was amazed to discover the glasses already in place. On inspection I found the second set to be ones I had lost over six months previously. It was as a result of their loss that I had purchased the new set little more than a week earlier. I had been struggling along without reading-glasses for about six months. My friends all laugh at this and offer somewhat insulting so-called "rational explanations". But I know none of these will fit.

My wife Margaret has experienced far stranger things than I have. The first happened during the war in the year 1944. A land-mine fell near the house in London and blew a heavy oak door in on her. She was only 10 years old and so the incident caused her much distress as well as some injury. She kept having nightmares and in one of them an explosion split the house in two down the middle and killed all the family. To give her a chance to recover her mother took her to stay with friends for a few weeks. They lived on a farm in Oxfordshire. On the day her mother was due to come to collect her she failed to turn up. Then the farmer's wife told her there had been an accident. The news was broken gently. Apparently a V2 rocket had fallen directly on to the shelters, killing all the residents of the entire street. When she saw the house it was split right down the middle, exactly as she had seen it in her dream. Her mother and two brothers were all dead, killed by almost the last V2 rocket to fall on London.

She has also had two "out of body" experiences when close to death's door. I will describe one of these. She was in a hospital bed following an operation and suddenly called the nurse. "I'm sorry but I have just wet the bed, couldn't help it." When the nurse threw back the sheets a spurt of blood shot up to the ceiling from a ruptured operation cut. She passed out through loss of blood. But she regained consciousness of a kind, viewing her unconscious form in the bed from a vantage point near the ceiling! She watched herself on the bed and saw everything which happened before feeling a warm glow bathed in a red light. She also tried to pull back the sheets but her hands seemed to pass through them, making no contact.

Now this is not an uncommon experience. In a television video, "Beyond and Back", a large number of similar cases are studied. They cannot all be fabrications. Why should they be? Peter and Mary Harrison, writers of many books relating to the paranormal, have made a deep study of the subject. They have told me that from their research more than ten percent of all people have been found to have had at least one such experience during their lives. For example, the descriptions given by people who have experienced the event are remarkably similar, regardless of country of birth or religious belief. The prospect of these being true records of things which have really occurred needs to be taken as a plausible alternative to the idea that they are all stuff and nonsense. We need to look for possible explanations based on extensions to physics.

Others have reported different phenomena for which no theoretical explanation yet exists. For centuries there have been reports of poltergeists, telepathy, clairvoyance, water divining, temporary materialisations of people once known to have lived and other so-called "paranormal" happenings. Some are no doubt fabrications but my own experiences lead me to discount the suggestion that all are explicable this way. Is the world crazy or is science lacking? Could it be that science has not developed sufficiently?

Not all scientists think in materialistic terms alone. The eminent Cambridge physicist, Professor Brian Josephson(116), published a paper, "Physics and Spirituality: the Next Grand Unification?" in 1989, saying that it ought to be possible to explain the

ONE WHO BELIEVES GOD CREATES A SOUL FOR EVERY HUMAN BEING BORN.

paranormal in terms of quantum physics. He made it
clear that science ought not to dismiss the paranormal.
Paul Davies(105) in "God and the New Physics" seems to
leave the question open. He does not attempt to support
the expectations the title suggests. My own impression
of this work is that he would like to give scientific
support to a creationist view but feels it prudent to
pull his punches. Why?

Some parapsychologists also support the creationist
view. There are two camps, however. One is purely
materialistic and seeks the so-called "rational"
explanation. This really means they do not accept any
view which reaches beyond the universe we can sense.
For example, they say that the "mind" can be explained
entirely in terms of functioning of the brain, that no
other organ needs to exist to explain the mind. This
may be true, but this does not exclude the possibility
that the mind may actually exist, built from some other
form of interpenetrating matter. In this event the
brain would be functioning as an interface system, so
that the mind could operate the body. It would be
similar to electronic interfaces used to enable
different types of computer to exchange information.
Unfortunately the establishment camp will not entertain
this alternative option and they hold all the aces when
it comes to the communication of ideas.

The other parapsychologists, those who take a
broader view and are prepared to probe deeper, are
consequently rarely heard. They accept that all may not
be confined to what we can sense. They look to
interpenetrating systems of matter, or, in usual
terminology, "parallel universes". The mind is not
excluded from residing as a separate entity in one of
these or even as a totally separate entity.

Rupert Sheldrake(121) talks about "morphic
resonance". He advances a convincing argument showing
that the first time anything is achieved, the difficulty
of achievement is always very great. The difficulties
always become less as the procedure is repeated, even if
no apparent communication has taken place. He cites

THAT WHICH OCCUPIES SPACE, AND WITH WHICH WE BECOME ACQUAINTED BY OUR BODILY SENSES: THAT OUT OF WHICH ANYTHING IS MADE.

substances whose formation into crystals has at first been found intractable and gives other examples. He concludes that a hidden factor has to be in existence which joins everything together. David Bohm(102) suggests that from quantum theory there must be some underlying unity in the universe.

Do these factors point to some sub-structure of the universe which connects everything to everything else? And if so could this be connected with psychic energy and/or a creationist scenario?

At this stage it is worth looking more closely at the ideas which are currently held by physicists regarding creation of the universe. According to classical physics energy can neither be created nor destroyed. So states the "First Law of Thermodynamics", which has been accepted as an absolute law of nature since the time of Newton. Quantum theory accepts this as true but only on time-scales which are large as compared with those of certain quantum effects. On minute time-scales it is thought that energy can arise from nothing as if borrowed temporarily from nothingness. It is a debt needing to be repaid an instant later. Hence a so-called "virtual energy" can exist temporarily as a quantum fluctuation arising from nothing.

According to "Big Bang" theory a huge quantum fluctuation arose by accident. The bigger the fluctuation, the shorter the time it can exist. However, another effect arose in this brief period. The space containing this spontaneous surge of energy developed a "negative pressure" and from this assumption it can be shown that all the energy needed to make both matter and its motion could have been scraped up from nothingness. A huge repulsive force arose together with this negative pressure, like greatly amplified gravitation with the direction of force reversed. A violent expansion resulted as the matter of the universe arose, each particle flying outward from the centre at speeds close to that of light. It does not matter if this seems incomprehensible. It is simply a brief summary of current thinking as described more fully by Tryon(123). Established theory maintains that this strange happening occurred during a short period called the "inflationary phase", after which no more energy was created. Most of it was in the form of light but a relatively tiny proportion existed as a mixture of hydrogen and helium gas. These gases were the primordial substance of the embryo universe. Then the whole universe continued to expand against the now universally attractive force of gravitation, like a growing balloon. Matter appeared as a rapidly expanding cloud of gas gradually slowing, eventually to a possible stop. In the meantime gravitational instabilities caused the gas to clump into local clouds the size of galaxies. Then smaller-scale instabilities, arising within these proto-galaxies produced the first generation of stars.

The larger a star, the shorter its life. Some huge stars ended theirs at an early time by exploding as supernovas to spread their contents over vast volumes of space. This was an essential first stage for the creation of Earth-like planets because only during such explosions could the heavier elements, such as iron and those needed to form rocks, be produced. They had to be created from lighter elements by nuclear fusion and only supernovas could generate the extreme pressures and temperatures required.

This enabled a second generation of stars to condense from the primordial clouds of hydrogen and helium with a little of the extras mixed in. During the initial phases a huge disc of rotating gas would form. The centre part would then form the star and this would begin to generate energy by nuclear fusion and radiate it as light and heat as soon as its density became sufficiently high. The rest of the disc would form into orbiting planets. Giant gas planets formed at great distances, but inside their orbits a few solid planets formed, consisting mainly of the extras.

Some of these planets were just the right distance from a star to create temperatures suitable for the evolution of life. They also had just the right mass for retaining an atmosphere. Many other conditions had to be just right by pure accident. Then further accidents resulted in the life-forms we observe, following the evolutionary process described by Charles Darwin.

This seems to be an acceptable description of the way the universe arose in broad outline. What appear to have been accidents could, however, be equally attributed to deliberate intention. We shall explore this possible alternative.

It may seem strange to lump matter and motion together but Einstein(110) derived his famous equation, $E = m.c^2$, at the beginning of this century, showing that energy "E" was equivalent to mass "m" multiplied by the square of the speed of light "c". Matter can therefore be considered as made from either mass or energy, since the two are equivalent.

A LUMP OF MATTER

In earlier times the amount of matter of which an object consisted was represented by its "mass". This could be determined, for example, by the act of weighing the object. Einstein's revelation still allows this but in addition permits the result to be expressed in the units of energy. Classical Newtonian mechanics, or physics, call it whichever seems best, already showed that energy is required to accelerate an object to a higher speed. Some of the energy contained, for example in a fuel, needs to be converted by an engine into the "kinetic" energy of motion of a car as its speed is increased from a state of rest. So energy exists in the fuel as "chemical" energy. Indeed energy can exist in many different forms, inclusive of heat, electrical and magnetic kinds. And all these forms can in principle be converted from one to another. It is even possible to convert matter into energy and back again. Consequently matter is said to consist of "mass-energy". All these forms can coexist in an object and can be added together to yield a "total energy". For example, the total energy of a moving car would consist of both its mass and kinetic energy. The chemical energy of fuel in its tank could only be released by combustion with oxygen and so is a complication which is best not considered here. All such forms of physical energy can be treated as interchangeable because they can all be measured in the same units. When at rest, an object can therefore be imagined to consist of "rest energy" alone.

These are forms of physical energy, however. They could also be thought of as "real" energy because it is possible to measure the quantities involved in physical units. Indeed any form of physical energy can be measured in units known as the "joule". This can tell us how many units of energy are needed to boil a kettle of water and the same units can be used to tell us the amount to expend in lifting a weight. Indeed all chemical processes and all electrical or mechanical forms can be determined in a very precise way by measuring them in the common unit of the joule. Yet at a deep level it is impossible to imagine the true nature of physical energy.

The astute reader may have already spotted a contradiction. No matter how all the energy was scraped up from nothingness in the Big Bang, overall everything appeared from nothing. This was a massive violation of the first law of thermodynamics, one of the two most basic conservation laws of this discipline. It simply states "Energy can be neither created nor destroyed". Yet after the inflationary period it is asserted that this law is obeyed exactly at all times and forever. It will be necessary to resolve this contradiction within our new theory.

Physicists ignore psychic energies entirely. They do not officially admit to their existence, though some will express their doubts on this matter in private. It has not yet seemed to fit into their domain. We will be looking into this, however, and we are going to find that it seems to have impinged already. As a consequence the reader will be asked to judge whether or not the facts have so far been misinterpreted. Have theoreticians been thrown on to a false trail in consequence? If physical and psychic forms of energy coexist, then in some way the universe must depend on both, not just on the one most easily measured. If both forms are used together to structure the universe, then does this mean that existence has an ultimate purpose?

The established view of physicists gives a clear "NO" to this question, because in their view psychic energy does not exist. They therefore support the view that all that "is" arose by happy accident. But are they correct? Popularisations of science, like those of Paul Davies(104 to 108), show that the odds against a universe arising with all its properties and laws of physics arranged such that life-forms could develop, are astronomically large. The figure one part in ten to the power forty, which means a one with forty noughts after it, is quoted for not just one property but for each of a whole array of properties as the chance of creating a universe like ours, capable of supporting life. Then this is multiplied by itself as each new property is considered. For example, with four such properties the chances against become one in ten to the power 160. Would you be prepared to back a horse if its chances of winning were one with one hundred and sixty noughts after it *against*? Yet this is precisely what cosmologists are doing when they look for a purely accidental explanation for the origin of the universe together with forms of life! A purely accidental origin seems unlikely, to say the least.

Why is science still treading a path which its own predictions show to be a very unlikely option? Should it not at the same time evaluate other alternatives? It should be an axiom of good science that all possible alternative solutions to any problem be explored simultaneously until any one meets an insurmountable barrier. This is prudent, because until a solution is established beyond any shadow of doubt it is impossible to know that the selected favourite was the right one to back. If the wrong choice has been made initially, then science can be thrown badly off course. It could become stuck on a false trail, struggling forever.

Yet what happens in practice is that in any one

quest a single "establishment" line always develops. Assessors responsible for the selection of material for publication toe the selected line. Then other ideas are ruthlessly rejected. Even when the established option has clearly failed after decades of futile searching, it is impossible to communicate alternatives. This happens even when an alternative offers a complete solution. It happens **especially** if an alternative offers a complete solution, because it is then perceived as an even bigger threat to the established approach!

This scenario has been repeated with depressing regularity throughout the history of science. It is being repeated again in the search for a viable solution to the problem of quantum gravitation. Here the established line is to try and match quantum theory to Einstein's theory of relativity. He first produced a version called "Special Relativity". This considered objects moving at speeds comparable with that of light and free from the influence of gravity. Later he extended the theory to his, "General Relativity", which provided an explanation for gravitation by his extraordinary concept of "curved spacetime". No force is needed to cause matter to fall according to this idea. Instead objects take the easiest route; they follow "geodesics" which can look curved to people like us but will actually be straight when observed from the vantage point of curved spacetime.

Essentially relativity is a physics of the large scale: it describes the motion of stars and planets for example. Quantum theory on the other hand attempts to explain effects on the very small scale: the sub-atomic range of size.

Quantum theory has a totally different basis from relativity. In quantum theory energy comes only in discrete chunks and requires forces pressing against objects to cause them to accelerate.

But even such eminent theoreticians as Stephen Hawking state that these two theories are now known to be incompatible. Yet they expect a solution will be found before the turn of the century by writing quantum theory in ever-increasing numbers of dimensions to make it fit relativity. The higher dimensions postulated have greatly increased curvatures of their space. So it is considered that the strange geometries involved can account for the other forces of nature in a manner similar to that of Einstein's geodesics in curved spacetime.

Yet after more than fifty years of effort by several generations of theoreticians they still think they need another ten years! Is it not therefore prudent to look at an alternative? This will start without Einstein's assumptions or any of his predictions. It will start out quantum-based and yet build upon the physics which Newton developed.

Einstein himself would have supported the simultaneous development of alternative theories. According to a private communication from an American scientist, George Meek, Einstein said on his seventieth birthday:-

"Now you think that I am looking at my life's work with calm satisfaction. But, on closer look, it is quite different. There is not a single concept of which I am convinced that it will stand firm and I am not sure if I was on the right track after all."

Einstein was a wonderful person with immense imagination and showed concern about the way scientific discoveries were put to use. If he thought he had inadvertently thrown science on to a false trail he would be most upset. If time had been available he would have done everything possible to rectify the situation.

As will be shown, gravitation has a major influence in our search for a viable solution to the meaning for existence. Those features which affect the main theme are summarised in the non-technical part of this book. But gravitation is so important a component that an entire technical section is devoted to presentation of the new solution. In this way it is hoped that

confidence in the approach will be communicated. This
is a simplified version so written that only "A"-level
school mathematics is required for its comprehension.
The first chapter (T.S.1) is a written version of the
lecture presented to a packed audience at the University
of Leeds on 18th January 1990.

The lay reader will be able to judge the issue by
comparing a summary of the two approaches. It is
sufficient to study only the inputs and outputs. These
are the initial assumptions and the results of checks
with experimental observation, taking into account any
unsolved problems or internal contradictions which
arise. A list of assumptions is given in CHAPTER T.S.1
of the TECHNICAL SUPPLEMENT. TABLE T11 is included at
the end of CHAPTER 11 and lists the results of attempts
to match Einstein's general relativity to quantum
theory. The same table shows how the new extended
Newtonian physics, which starts out quantum-based,
satisfies the checks just as well as general relativity,
yet is totally free from internal contradiction and all
the other difficulties which confound the established
approach. It also relates the magnitude of the
gravitational force to that of electromagnetism, which
neither quantum theory nor general relativity have been
able to achieve.

The established scientific approach regarding the
origin of the universe is based on accidental formation
followed by the accidental arrival of all life-forms
including ourselves. This is despite the slender
chances which are freely admitted. Yet as will be
shown, theories which provide plausible solutions to the
dilemmas of gravitation and wave-particle duality lead
to different conclusions. Is it therefore just possible
that establishment science could be wrong in some key
areas?

But this science has already taught us many useful
things. It has taught us to beware of illusions. Our
senses often deceive. The ancients saw the dome of the
sky studded with points of light they called stars. The
Earth was taken to be flat, forming the base of the sky-

dome. Then both the moon and the Sun scudded across the surface of this dome every day. Somehow, they speculated, something dragged them back under the Earth so that they could reappear in the correct places to begin the next day. Then their astronomers noticed some wandering stars which moved independently of the majority which were fixed. Finally they conceived of a universe centred on the Earth with the sky rotating around it as a hollow sphere studded with stars. Then mathematicians found rules to predict the motion of the Sun and moon and all the wanderers - the planets, of course. This was the elaborate Ptolemaic system, very clever in its way with its "wheels within wheels" to describe the apparent cycloidal motions of the planets. This theory was made ever more complicated and ungainly as it attempted to explain more and more new observations. Eventually it became impossibly unweildy. Is history repeating itself?

The model simply assumed that everything was exactly what it seemed to be, a childlike interpretation which made no allowance for the limitations of our senses. People did not appreciate that our eyes can only assess distances up to a quite limited range. Beyond that all objects appear the same distance away. Hence the Sun, the moon and all the stars appear exactly the same distance from us. This distance corresponds with the apparent radius of the sky-dome. To compound the problems the nature of gravitation was not appreciated. The Earth could not be round, even though the horizon of the sea seemed to suggest such a shape, because you would fall off on the other side. The ancients did not realise that the force of gravitation meant being pulled toward the centre of the Earth. "Down" means on a radial line drawn toward the centre of the Earth and so is not parallel to directions measured at other places. These two factors together caused completely false interpretations to arise. Similar misinterpretations can still happen in our time. We must beware of them.

Later Copernicus simplified the system by suggesting that the Sun, not the Earth, formed the centre about

which the universe revolved. Suddenly all the
complexity fell away! He was careful not to publish his
ideas until he lay on his deathbed, however. In those
days ridicule was not the greatest danger faced by the
seeker for the truth.

The trouble was that primitive interpretations had
been incorporated into the Christian doctrines. The
latter were based on faith alone. Once inserted, any
progress in understanding which suggested flaws existed
appeared as a threat. Faith could not survive if part
of the doctrine needed revision. The collapse of a part
could undermine confidence in the rest.

Then when Galileo produced a telescope which proved
that moons could revolve around Jupiter, just like a
model of the solar system described by Copernicus, the
fat was truly in the fire. The solution was to refuse
to look in the telescope. True, Galileo managed to get
an astronomer into his observatory. But that was the
nearest he came to viewing. His reason?

"I do not need to look in the telescope because I
know the planets do not exist. If they existed they
would be visible to the naked eye, because if they are
not visible they can be of no use. Therefore they
cannot exist."

We laugh at this now. What many of us do not
appreciate is that exactly the same techniques are still
used to discredit truths which appear as threats to the
establishment!

As is well known, Galileo was forced to recant by
the notorious Inquisition. The progress of science was
delayed but it was not halted. It grew estranged from
religious doctrine and in the end establishment theology
was the loser. For centuries the advance of science
emphasised materialistic solutions to the puzzles of
nature. It seemed to leave no room for religious
interpretations and the authority of the Church
gradually eroded away. Today fewer than two and a half
per cent of the people in Britain regularly attend

church. Its doctrines seem too implausible. They seem absurd when set against the proven knowledge of science. The result is that most people, if not atheists, are agnostics, careful not to probe or think too much about these matters. Yet as this book unfolds it will become evident that the two basic religious concepts cannot be destroyed by science. These are the idea of an intelligence which structures matter and the idea that people can have consciousness separate from their bodies -their souls.

Because the Church refused to adapt, a wedge of antipathy was driven between science and theology. Science is now constrained by its own momentum to avoid supporting any "creationist" view of the universe. It is professional suicide to suggest in any scientific paper that the universe might have arisen at the will of a super-intelligence. Hence, even though many scientists secretly hold the view that this is more likely than an accidental origin, they dare not admit to holding such opinion in public.

However, everything arose either by accident or as a planned creative act. There are two options and it is therefore in the interests of good science that both options be explored with equal rigour until one of them meets up against an insurmountable barrier. The symptoms arising when such a barrier is met manifest themselves by a plethora of fanciful notions as every straw is clutched. Some very fanciful ideas have been appearing for some time as the accidental option of the established route is pursued. The physicists view of the universe has again become hopelessly complex and implausible. Worse, it contains a number of internal contradictions and even one is inadmissible. It is worth reading an article by Abbot(101) in order to arrive at some judgement of this issue. Is it not time, therefore, to look at the alternative option?

Perhaps things have been missed which could extend science so as to encompass a wider range of nature's secrets. Then with luck some of the conflicts in present theory may be resolved, so providing a new

simplified framework to explain the universe. We shall
explore this suggestion as the book unfolds.

1.2 KEEPING AN OPEN MIND

For those having a technical background some parts
of the new theory may seem absurd or ridiculous at first
sight.   It is particularly important that one very
plausible reason for this response be considered. This
is because the mind is prone to respond in such a way to
new ideas. If history is followed it will be observed
that whenever any basically new concept has been
introduced it has been met with ridicule and abuse.
This has often delayed acceptance of the truth. Yet,
years later, people are amazed when they look back and
recall the almost universal initial response. From the
later vantage-point the initial response always seems
unbelievably stupid.

Why does this happen? The answer is that one of the
hardest things to do is to maintain an open mind. This
is because almost everybody thinks they have one. When
did you last meet a person who claimed to have a closed
mind? In fact there are quite a lot of open-minded folk
but they are nearly all under the age of twenty-one.
After about this age, when the subconscious mind judges
that all the concepts needed to see life through have
been absorbed, it clams tight shut. But a major defence
strategy of the subconscious is to deceive its owner
that it is always open. No wonder nobody ever claims to
have a closed mind!

I read an interesting article written by a doctor
when I was about 22 years old. Unfortunately I kept no
record of the reference but the substance of the article
made a big impression. So I have always retained it
very clearly. He said that a part of the brain, which
I think he called the hypothalamus, acted as a filter to
remove unacceptable material. Incoming new concepts
delivered via the senses were first routed to this part
of the brain for checking. They were compared with
existing concepts stored away in the memory to see if
the new ones were compatible with them. If not, then

they were rejected. The point is that incompatibility was equated with the new concepts being wrong.

Now in any branch of science some concepts are always bound to be false, otherwise there would be no need for progress. One has therefore always to be on the lookout for new concepts which could be better than the old they are attempting to replace. But this is easier said than done.

The mind rejection mechanism acts rapidly, so rapidly that the logic centres are bypassed. Instead the emotional centres are activated. Then the immediate response to an incoming idea which has been classed as incompatible is rejection. This happens before it can be analysed. The rejection mechanism activates the emotional responses of hostility and absurdity. A desire to pour ridicule on the new is experienced. There is, however, no justifiable reason at all at this stage to classify the new as false.

The way to keep an open mind, therefore, is to recognise the symptoms of the closed mind. It is necessary to make a deliberate effort to keep an open mind. It will not stay open by itself. One needs to recognise this without being ashamed to admit the fact. Then a strategy can be adopted to counter the difficulty. My own strategy seems to work for me and so I will now try to explain how I deal with the problem.

As soon as I experience the feelings mentioned, a symbol **"MBM"** flashes in bright red letters in my mind. It means **"Mental Blockage Mechanism"**. It is a warning I have trained myself to experience. Then instead of following my instincts I can then look at the issue in a rational way without discarding it out of hand. Very often I find ideas, even those of my own which I would normally have rejected, become acceptable because I cannot find a logical flaw.

This is one aspect but there is also another. Most people make all-or-nothing decisions when faced with a new choice. This is necessary in a business environment when a practical decision needs to be made, but there is no need to do this when assessing new ideas. It is particularly necessary to avoid making a choice when the available information is inadequate. My method is to visualise a 0 to 100 scale. Then a subjective rating is given. A zero rating means total disbelief with rejection and 100 total acceptance. Rarely can the terminal values be allocated because these are only allowed when absolute proof is available. On this basis, for example, nobody could give a zero rating to the possible existence of the "soul", because it is impossible to *prove* it does not exist. Therefore if a low rating of only say 10 is given, the matter is kept in mind and not rejected. Then as more information comes in to improve the certainty rating in either direction, the rating can be modified. Ultimately complete rejection or acceptance can become justifiable.

It would help if children were taught at school to assess this way. Then when they become adults, many of the barriers still delaying progress will at least be made easier to surmount. For similar reasons it is important to avoid the implantation of false concepts at an impressionable age.

With these thoughts in mind we can now turn to our main quest, the search for a satisfactory extension to physics, able to provide a framework capable of integrating physical and psychic forms of energy. The extension needs to be capable of admitting and explaining all known data of the so-called paranormal kind. At the same time a solution to the vexing question of explaining gravitation in terms of quantum theory is to be sought. This will need to satisfy all experimental checks just as well as does Einstein's theory of general relativity but at the same time it must be free of its difficulties and fit in with the paranormal section. It must also be consistent with quantum explanations of the other three forces of nature, namely the strong nuclear force which holds the nuclei of atoms together, the weak force responsible for the radioactive decay of heavy nuclei and the force of electromagnetism. In a short book only certain data can be selected from the vast amount available, but the result needs to be compatible with any other which can be used.

Data from the experiments of physicists relating to the universe which we sense is to be mixed with so-called paranormal inputs. The data from astronomical observations and the laboratory regarding the strange nature of light and matter, of the dynamics of energy, the mechanism of electric charge and gravitation, are to be mixed with data concerning psychic energies or forces.

No attention will be paid to the existing taboos which have been the unfortunate legacy of history. We shall cut across interdisciplinary boundaries to relate factors normally thought of as separate issues. Factors which are put in separate boxes in the conventional treatment. But everything is connected to everything else. If things are treated separately for convenience, then the connecting links between boxes must not be ignored, since to do so can lead to false conclusions.

At this stage the complexity of the problem we have set ourselves may seem overwhelming. We have asked many imponderable questions and have as yet sought no answers. In fact things are not so difficult as they may appear at first sight to be. It is necessary, however, to try and keep all the elements of the problem at the back of the mind, whilst seeking to comprehend each in turn. Surprisingly, when this is done one problem seems to help out the others in a cross-fertilizing manner.

The problem of wave-particle duality is one of the two key factors, as will be shown. This will link the physical to the psychic forms of energy. The other key factor is quantum gravitation. It will lead us to new insights regarding the physical kinds of energy. These two factors are inextricably linked and together will lead to exciting new horizons. They will lead to the deduction of a computer-like grid structure permeating the entire universe which organises matter and the motion of matter. They will support the idea of a Creator which not only started the universe on its course but needs to be there all the time to maintain its structure.

The new concepts of gravitation will be explained. To support them a technical supplement is provided. In this a new range of experiments is described. It is my hope that this book will trigger in some fresh and active mind the enthusiasm by which some of these experiments will be tried out. One of them seems especially exciting. It shows that an entirely new and totally non-polluting source of energy might just be waiting in the wings.

2 HISTORICAL BACKGROUND.

THE PROBLEM OF WAVE-PARTICLE DUALITY ARISES

The problem of wave-particle duality is rooted deep in history. Even during the lifetimes of Galileo and Newton conjecture arose concerning the nature of light. Did it propagate as a system of waves or did it consist of a stream of particles? Even today no solution has been established.

The history is in itself extremely interesting and so this chapter concentrates on its development. At the same time all the essential knowledge of physics, needed for a total appreciation of the climax of our theme, should be imparted to any intelligent lay person. We cannot deal with just light alone because other factors are involved, such as some of the basic laws of physics.

By presenting the subject matter as a brief historical account, however, it is hoped that a painless introduction to a fascinating subject will be provided. Then the reader will be able to appreciate the difficulties of interpretation facing physicists. The data seems crazy and conflicting. It will be shown that even today none of the solutions to the problems raised by wave-particle duality, which have been tentatively accepted by physicists, are really satisfactory. And everybody knows this, so the search continues.

What is thought to be a new and novel solution is described in later chapters. The description is purely verbal, no mathematics is needed. Yet it is shown

capable of satisfactorily relating known data to a
remarkable degree. It is for the reader to judge its
success. The solution started out following established
lines with the assumption of an accidental origin of the
universe in mind. This became more and more untenable
as a solution emerged. Finally the solution described
homes in on a creationist scenario which appears to
justify the faiths people have relied upon throughout
the ages. This also answers the oft-posed question,
"How could God have made himself?" It shows that nature
allows options supporting the idea that people could
have souls. Physics cannot prove they have them but
experience has shown that nature abhors a vacuum. All
possible niches seem to be filled.

2.1 READING PLANS

The reader having no technical or scientific
background is likely to find some of the concepts
introduced difficult to comprehend at the first reading.
If this is the case then a more leisurely study is
recommended. The difficulties being encountered
probably arise from a too rapid introduction to one new
concept after another; it takes time for the mind to
accept them. None are really difficult though they may
appear so at first. It is best to read a little and
then come back to it later, giving the subconscious mind
time to digest. The mind often seems to act like a
separate organ, sometimes acting contrary to our wishes.
It is necessary to learn how to trick it into doing what
we want. If interest is found to wane a little then a
quick look at bits of Chapters 9, or 13, followed by a
rest period, should do the trick.

These two chapters, however, only show how the new
theory can explain psychic phenomena and spirituality in
terms of the "Grid", a construct developed by the new
theory. It is a computer-like structure pervading all
matter and space to interconnect everything to
everything else. It provides the invisible base on
which matter is organised so that the illusion of a
solid world is created. This is needed for finding a

plausible solution to the dilemmas posed by wave-particle duality and also for resolving a contradiction in explanations for the shapes of atoms. For a brief view of this without too much detail, Chapter 8, which summarises the solution, should be found helpful and, perhaps, more readily digestible.

Alternatively one or two of the very simple experiments described in this chapter can be tried. Each apparatus is quickly and easily constructed and set up. Science develops with experiment and theory advancing together hand in hand. Experiment helps new concepts to gel in the mind.

For those who find they can relate without excessive effort to the logic of the scientific way of thinking, the best approach is to read straight through. This way should convey the deepest understanding of the new extension to physics which I am trying to communicate.

2.2 GALILEO

Galileo Galilei, according to Shamos(120), was born the same year as Shakespeare, in 1564, and is the recognised father of the scientific age. The ancient Greeks had worked in abstract ways to develop mathematics and were very successful in their efforts. They did not, however, attempt to use it together with practical experiment to provide a true science. This was the step which Galileo made. He rolled cannon balls down inclines to measure the rate at which their speeds increased. In other words he measured their "acceleration" and formulated theories which enabled the mechanics involved to be understood. He found, for example, that however massive the ball was, on a given incline the acceleration was always the same. Yet the heavier the ball the greater would be the force or distance needed to bring it to rest after it ran on to the level. To obtain a single quantity which would specify the effort needed to bring any ball to a stop he multiplied the mass, found by weighing, by the velocity and called the product "momentum".

This is a very important property in all mechanics.

Both the velocity and mass contributed to the effort required to stop any object. Any object showed a reluctance to change its motion. If at rest a force needed to be applied to start it moving; when in motion a force of some kind was required to stop it moving. In other words it possessed "inertia" and momentum provided a measure of this property. Momentum will come up again quite frequently as our story unfolds.

He also experimented with pendulums. Indeed a surprising degree of insight into the mechanics governing our universe can be gleaned by studying the simple pendulum. He discovered that, provided the weight of the thread was negligible in comparison with the compact pendulum bob it supported, the time required to complete one swing was the same no matter how heavy or from what material the bob was made. The period was also unaffected by the angle of swing up to about a 30 degree inclination to the vertical.

This so-called "periodic time" only depended on the distance between the centre of the bob and the point of suspension. To obtain a periodic time of one second, for example, this distance needs to be 24.8 centimetres. To treble this period the length needs to be nine times as great. In general the length required needs to be increased in proportion to the square of the periodic time specified.

Then he tried placing round pegs below the suspension point so that a new pivot was created at the half-swing position. The second half-period was then greatly reduced as compared with the first and the angle of swing increased. The most interesting observation he made, however, was that when the bob rose to the point where momentarily it came to rest, it always reached the same height, above a level floor, as the height from which it was released.

This may not at first seem very remarkable but, nevertheless, the implications are profound. The interpretation accepted, even today, is that two different forms of energy are involved which interchange with one another. As the bob swings down its speed increases and associated with this speed is an energy of motion called "kinetic energy", which we will subsequently denote by the symbol "KE". Because this energy has to come from somewhere it is assumed to derive from reduction in height under the action of gravity. At a high level the bob will possess more so-called "gravitational potential energy" (GPE) than at a lower level.

An arbitrary level can be chosen as the datum from which measurement is to be made. This could be conveniently chosen as the lowest point reached by the bob. Then at the highest point, where the speed falls to zero, all the kinetic energy will be recovered as an increment of GPE. When it swings back to the lowest point again all the increment of GPE stored will be converted back again to KE.

With this explanation even though symmetry of the oscillation has been destroyed by adding a peg, the bob must rise by the same height increment in order to convert a given amount of KE to GPE.

Furthermore an oscillation consists of repeated interchange between GPE and KE and could go on forever in principle. Galileo used pendulums as timing devices for his experiments but the amplitude of the swings always decayed to an eventual stop. This was caused by parasitic effects such as air resistance which absorbed a little of the energy at each swing. To offset this loss it was necessary to give the bob a little push at the start of each period, a push carefully timed to match the natural period. This is what the "clockwork" mechanism ultimately achieved.

These experiments of Galileo are well worth trying and are easily performed.

He made other important contributions we will not mention. Unfortunately his astronomical observations supported the theory of Copernicus (1473-1543). This was the "heliocentric theory", which displaced the Earth from its special position as the centre about which the Sun and indeed the whole universe was believed to revolve. This was perceived as a threat by the

established Church because it ran counter to their
teachings. As everyone knows, Galileo was forced to
recant by the Inquisition. So conflict between science
and religion arose right at the beginning of the age of
science. Only now, late in the 20th century, does the
rift promise to heal.

Wave-particle duality, one of the two main topics of
this book, is the key factor in this new turn of events,
as will be shown. It arose when somebody else began
studying the behaviour of light.

A painter 54 years younger than Galileo worked for
the Pope and must also have been a gifted person. He
was Francesco Grimaldi(120). He experimented with light
by allowing it to pass through small apertures and
discovered some interesting effects. It is very easy
and informative to try and reconstruct some of the tests
he made. He used the Sun of course, but it is easier to
use a slide projector. This can be set to focus in the
far distance to obtain a parallel beam. Then pieces of
thin cardboard can be used as masks to partially
obstruct the beam, which is finally allowed to fall upon
a white screen about two metres away. If a slit about
3mm wide (1/8") is made by a sharp knife, then most
people would expect to see a thick bright band of light
displayed on the screen. Instead, what is actually seen
is a whole array of light and dark bands, together with
some rainbow colours, covering the screen. Now a second
piece of card with a similar slit cut in it can be held
vertical and a few inches behind the first. The card
can be rotated so that the new slit is inclined at about
45 degrees to the first slit. Then a cross-hatched
pattern covering an oval patch of light will appear.
These tests are very interesting and are well worth
trying.

To Francesco this seemed analogous to the
"interference patterns" which can arise on a pond.
These occur when two pebbles are thrown in together.
The subsequent ripples mutually interfere in the region
where they cross over one another. He deduced from this
that light must have a wave nature. When one realises

the state of knowledge at the time, this was a
remarkable deduction. Later both Christian Huygens
(1629-1695) and Robert Hooke (1635-1703) supported this
conclusion. Unfortunately they were contemporaries of
Sir Isaac Newton, who produced his "corpuscular theory"
for light. He imagined light to propagate rather after
the fashion of a stream of tiny machine-gun bullets.
Each was a separate "corpuscle". Nowadays these are
called "particles".

2.3 NEWTON'S PHYSICS

Newton (1642-1727), probably the greatest genius of
all time, is famous for creating a whole new system of
physics which still forms the basis of most scientific
and engineering calculations. He developed a new form
of mathematics, his "calculus", and used it to explore
his new ideas of mechanics. He defined what energy is
and showed this to be quite different from the concept
of momentum.

He caused balls to bump into other balls and
measured the speeds before and after collision. By
interpreting the results of such experiments he made a
startling discovery. A simple mathematical "law"
existed from which the outcome could be predicted. This
is called "The Law of Conservation of Momentum". He
found that if he added up the momentum (mass times
velocity remember) of two balls after collision this
would always be equal to the sum of the momenta the
balls had before collision. This applied if the balls
were constrained to move only along a single line with
one of them stationary before impact. It also applied
if both balls were moving before impact in the same
direction. If they moved in opposite directions before
or after impact, then the law worked provided the
velocity of one direction was regarded as negative with
respect to the other. The direction of motion was as
important as the speed.

This is where the term "velocity" differs in its
meaning from "speed". Speed and velocity are measured
in the same units such as miles per hour or metres per

second (m/s) but speed is called a "scalar" quantity because the direction of motion is not included. Velocity is a "vector" quantity, meaning that speed and direction are combined.

There is a more general meaning to change of velocity which applies where motion is not constrained to a straight line. The difference in meaning between speed and velocity can then be provided, for example, by imagining the effects of cornering in a car at high speed. The car moves round the corner at a constant speed but the velocity is continually changing because of the changing direction.

The rate of change of velocity precisely defines what is meant by "acceleration" according to Newton. He then showed that when a mass is accelerated a force is involved. When a car is accelerated forward, for example, the people inside are also accelerated and consequently feel a force pushing them as they press against the backs of their seats. Here, simply considering the speed increase alone is sufficient because direction is unchanging.

But when cornering at constant speed a side force is encountered. Again this force is associated with an acceleration but this time it is due to the rate of change in direction. To make an object move in a circular path it is necessary to provide a force of constant magnitude but always directed at right angles to the direction of motion, then no change of speed arises. An example is the swinging of an object on a string to make it travel in a circle. The string applies the necessary force. Of course, in the general case, both the speed and direction can be changing simultaneously.

Because of similar observations, the term "mass", which represents the amount of matter present in an object, was defined by Newton from the force required to provide a given acceleration, or:-

"*Mass is equal to the force applied divided by the acceleration produced*". This is one of his laws.

Newton also produced a theory of gravitation which

said that a force was somehow transmitted through empty space between any pair of objects. This force tried to pull them together. One object could be the Sun the other a planet. The force was proportional to the mass of one multiplied by that of the other and divided by the square of the distance separating them. This is the well known "inverse square law". It means, for example, that by trebling the separating distance the attractive force will be reduced to one ninth of its previous value. The objects would not fall into one another by the acceleration produced by this force if at the same time a sufficient "tangential" velocity existed. This is the velocity measured at right angles to the direction of force.

In special cases the result could be perfectly circular orbits. It will be remembered that a circular path is traced by an object which is acted on by a constant force always at right angles to the direction of motion. Then Newton showed, by clever mathematical analysis, that in general the orbits of planets would be perfect ellipses. These are oval shapes with the Sun displaced from the centre. The point of greatest distance from the Sun is called the "apogee" and here the orbital speed is a minimum. At closest approach, the "perigee", the speed is a maximum. For the ellipse, acceleration is due to both change of speed as well as direction.

For those who would like a little more insight about Newtons laws, an appendix is attached at the end of this chapter. It explains the differences in meaning between the concepts of kinetic energy and momentum and describes momentum conservation in two dimensions. It also describes another simple pendulum experiment the reader might like to try which demonstrates the conservation of momentum and energy.

More important for our problem of wave-particle duality is that light, which according to Newton is made from particles, would behave no differently from other objects. A beam of light from a star grazing the "limb" of the Sun (its edge) would be deflected by the Sun's gravitational pull. The star would appear to have been deflected from its normal position in the sky. No such deflection would arise if light propagated only as a system of waves.

So great was Newton's influence that his opinion was not officially questioned for more than a hundred years. But this was the start of the dilemma. Did light propagate as a system of waves or did it have a particle nature?

2.4 YOUNG'S TWO-SLIT EXPERIMENT

Thomas Young (1773-1829) answered this question in favour of waves, though not until 1803. He is famous for his "two-slit experiment". He was a physician, mostly an eye specialist, but made contributions to a surprisingly large number of other disciplines such as Egyptology, science and engineering. "Young's modulus", for example, is a key factor for engineers which enables the deflections of structures to be calculated.

His two-slit experiment showed light must have a wave nature, being propagated after the fashion of ripples on a pond, though in all three dimensions instead of the two which can specify the surface of a pond.

He used a beam of light from the sun, reflected by a mirror, to pass horizontally through a small hole in a blind and into a darkened room. Then he allowed it to pass through a pair of narrow slits arranged vertically side by side and therefore parallel to each other. Further on, a screen was arranged to display the resulting pattern. If most other people at the time had tried this, they would have expected to see a pair of bright lines on the screen. But not Thomas Young. He saw what he was expecting. But this consisted of a pattern of light and dark stripes aligned in the same direction as the slits and extending sideways indefinitely as it faded away. This "interference pattern" is illustrated in FIG.1 and can only be explained by the superimposition of wave trains. Instead of forming fairly sharp shadows, as would be expected from the stream of particles concept, the light spread out by "diffraction" at the edges of the slits, forming two sources of waves in phase with one another. This means that if the waves are pictured as ripples, then the maximum heights are reached at equal distances from the slits. At the centre of the screen both sets of waves arrive together so reinforcing one another to cause "constructive interference" and produce a bright band. These light and dark bands are only seen edge on in FIG.1, where the screen, shown at the top of the figure, intercepts the light.

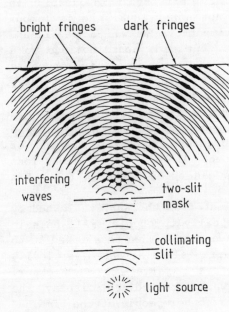

bright fringes   dark fringes

interfering
waves

two-slit
mask

collimating
slit

light source

FIG.1 YOUNG'S TWO-SLIT
EXPERIMENT

The waves arrive out of phase at points on the screen away from the centre and at a certain distance total "destructive interference" occurs. The crests of waves from one slit are neutralised by the arrival at the same instant of the troughs from the other. Hence a dark band is seen. Farther away still, the peaks of the nearest slit arrive a full wave cycle before those from the other and so peaks again coincide to produce another bright band or "fringe". The process repeats indefinitely, so that many alternating bright and dark fringes are formed. FIG.1 should be interpreted as an instantaneous frame in a sequence of frames in time, such as would be recorded on a cine film. All the waves shown are to be imagined as

progressing up the page and spreading outward until they
hit the screen. The places where the wave crests cross
each other can be seen. Here the maximum wave
amplitudes arise and to either side, for a substantial
distance, a high value is maintained. These regions are
marked by thick lines.

The pattern can be easily reproduced by a ripple
tank. A shallow tank containing water to a depth of
about an inch can be used with the slit masks arranged
in vertical planes fixed to the flat bottom. Slits
about 0.1" wide and separated by about 1" will be found
satisfactory. The ripples can be formed by making use
of the vibrations of a spin drier. A strip of metal
glued to the frame of the drier can be bent to dip into
the water at a place representing the light source. It
is important that the experimental tank has a mounting
separated from the spin drier. The screen needs to be
represented as a sloping beach to cause waves to break
and so prevent reflection.

Young's real experiment can be used to actually
measure the "wavelength" of light. This is a measure of
the distance between successive wave crests or,
alternatively, their troughs. A colour is selected by
a filter. This means that a certain wavelength has been
selected.   Then the distance to the screen, the
separation of the slits and the pitch of the fringes are
measured. By quite simple geometry the wavelength can
then be calculated. It is simply the centre to centre
distance between the bright fringes, multiplied by the
separating distance of the slits and divided by the
distance of slits to screen.

It has been found in this way that red light has a
wavelength of 7/10,000 of a millimetre whilst blue light
is shorter at 4/10,000 mm.   These values mark the
boundaries of human vision. But beyond vision at longer
wavelengths "heat rays" can be detected which are called
"infra-red". "Ultra-violet" rays, also invisible, can be
detected at wavelengths shorter than the blue. The heat
radiated by a hot object can be felt.   Then as the
temperature is increased it is seen to glow dull red as

well. Further heating causes the colour to change to yellow. A very hot object like a lightening flash is seen to be blue. By some strange quirk we, as people, relate blue to cold and red to hot. From the scientific viewpoint this is the wrong way round.

Although Young backed his experiments with a convincing theory, not much impact was made at the time. But ten years later Jean Fresnel, who is famous for his lighthouse lenses, produced an even better theory, showing how diffraction could be explained. This definitely established the wave nature of light.

Then Clerk Maxwell (1831-79) showed light to be electromagnetic in nature. He used abstract mathematical reasoning of a high order to produce "structural equations" but these were not physical structures able to be visualised. They did show, however, that light consisted of "transverse waves". This means a motion occurred at right angles to the direction of propagation, like that produced by the shaking of a skipping rope. They were a combination of electric and magnetic force fields moving in sympathy with one another. Light represented only a small part of the electromagnetic spectrum which extended beyond the infra-red to radio waves which could have wavelengths of many metres. Beyond the ultra-violet, radiation of very short wavelengths was discovered. These were the so-called X-rays, having great powers of penetration.

All these waves moved at the same speed, that of light which is still represented by the symbol "c". This had been measured by a contemporary of Newton, the astonomer Olaus Roemer, who gave a very accurate value. This speed is 300,000 km/s (kilometres per second). A few simple equations will now be quoted. If the reader feels put off by this in any way then a brief look at the appendix may help. It must be emphasised at this point that this book will not quote many equations and if they are ignored it will not really matter.

The frequency "ν" at which waves arrived at a given point was simply connected to the wavelength "λ"

by dividing "c" by "λ" i.e. "$\nu = c/\lambda$". If, for example, "λ" = 10m, then this equation shows the corresponding frequency to be 30 million cycles/s, which is usually stated as 30 megahertz. The frequency or, alternatively, the wavelength was all that differentiated light from other types of electromagnetic wave.

Many other famous people such as George Stokes and H.A. Lorentz added their contributions in support and so there seemed absolutely no doubt that light had purely a wave nature.

2.5 PLANCK AND EINSTEIN

Until Planck and Einstein arrived on the scene! The German physicist, Max Planck (1858-1947), had shown in the year 1900 that electromagnetic radiation had to be emitted in discrete packages of energy which could not be further sub-divided. These packages were called "quanta" and were larger the higher the frequency "ν" of the radiation. The energy value corresponding with each quantum of radiation is given by "h.ν", where, appropriately enough, "h" is known as "Planck's constant". It appears repeatedly throughout the whole of quantum physics. A beam of light would carry a total energy equal to the number of quanta transmitted multiplied by the energy carried by each. The rate of energy falling on every square metre of a surface defined the "intensity" of light and this would be increased by either increasing the number of quanta concentrated in the beam or by increasing their frequency. The puzzle is how to relate a frequency to particles. Later an answer will emerge.

Then in 1905 Albert Einstein (1879-1955) investigated the "photoelectric" effect. He called each quantum of light a "photon". Each penetrated the surface of a metal and gave up all its energy to a single "electron". At this point it is necessary to give some idea of the way atoms are structured.

2.6 THE ATOM

The ancient Greeks defined the atom as the smallest part into which matter could be subdivided. These were thought of as little hard balls. An element such as iron or sulphur can be divided to leave such isolated atoms, but although they seem in many ways to be solid and hard, what we sense is far from a true description of reality.

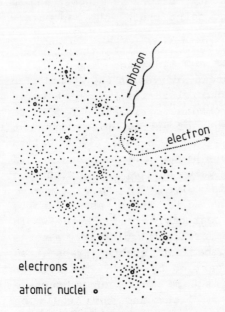

electrons

atomic nuclei o

FIG. 2 CUBIC CRYSTAL LATTICE
electron clouds of atoms

In fact, as shown by the scattering experiments of Sir Ernest Rutherford (1871-1937), they consist of a nucleus occupying only a minute fraction of the total volume, yet accounting for nearly all the mass of the atom, surrounded by even smaller and lighter particles, the electrons. The latter were discovered by J.J. Thomson in 1897. They are credited with "negative electric charge" and possess a total which is balanced by an equal amount of positive charge in the nucleus. There is a force of attraction between negative and positive charge and it is this which textbooks say cause the electrons to be held in so-called "orbitals" close to the nucleus. It is the orbitals, the space in which the electrons are constrained to move, which defines the size of the atom.

They appear hard because the orbitals of other atoms contain electrons also and these exert forces of repulsion upon one another. This prevents the orbitals of adjacent atoms from inter-penetrating to more than a very limited degree, even when pressed against one another with great force. There is a limiting force, then the atoms collapse to provide the very high densities of the "white dwarf star". When an ordinary star, like our familiar Sun, has burned its nuclear fuel, it first expands to a "red giant" and then by losing heat, contracts to a white dwarf. The atoms are crushed by immense gravitational force, now that the radiation pressure caused by light trying to escape is no longer present in sufficient amount to offset gravity. It now has the mass of the Sun compressed into a ball no bigger than the Earth. Yet the mass of the Sun is 333,400 times the Earth's mass!

Even this does not represent the ultimate density of matter. Larger stars sometimes explode into supernovae and their centres are imploded to become neutron stars. Then one solar mass is contained in a ball only a few kilometres across. This density corresponds with that of the atomic nucleus. But even this is not the ultimate density because the nucleons are themselves made up of tiny components occupying only a small fraction of the available space.

Hence electrons are sub-atomic particles having unit "negative electric charge". This charge causes mutual repulsion with other electrons and a flow of such charges constitutes an electric current. They react by a force of attraction to any positive electric charge. Such charges are carried by the "protons" which form part of the atomic nucleus. Electrons are arranged in cloud-like form defining the orbital about the relatively massive atomic nucleus.

The electron clouds of adjacent atoms stick to one another by short-range residual electromagnetic forces. Magnetic forces result from electric charge in motion. The electrons have both a spinning and an orbital motion, so that magnetism is generated. Then other electrons respond to this magnetism as well as the electric force. In materials like iron the magnetic force builds up from atom to atom and then becomes very evident.

The combination of electric and magnetic forces holds atoms together to make apparently solid structures or liquids.

In FIG.2 a few atoms are illustrated forming a crystal having a cubic lattice. Electron clouds, confined within volumes of space called "orbitals" about their nuclei are shown. The clouds around adjacent nuclei merge together to create the illusion of a solid structure. Solid to observers like us, that is, who are relatively huge in size. But this is simply because our fingers are also made of atoms whose electron clouds prevent interpenetration with atoms of materials which they are pressed against.

Einstein showed that the quanta of light called "photons" were absorbed by electrons, so giving up energy to raise the electrons to "excited states", where they existed for a short time at a greater distance from the nucleus. If the photon had sufficient energy it would knock an electron right out of its orbital. Since each electron possesses a negative electric charge and charge in motion constitutes electric current, light having sufficiently energetic photons would generate electric power. This is the "photoelectric" effect, also illustrated in FIG.2 by a photon knocking an electron out of the lattice. It can only be explained by light behaving as a stream of particles, because waves would have their energy spread over too wide an area. The energy in waves would be insufficiently concentrated to provide the observed effect. This could only occur by a photon acting like a billiard ball, hitting another and so causing a transfer of all its energy.

A simple experiment can be performed to illustrate electric forces. A sheet of thin brown wrapping paper about A4 size is first heated in front of a fire. This drives out any moisture which would spoil its electrical insulating properties. Then the sheet is pulled rapidly between the sleeve of one arm and the body to charge it with "frictional static electricity". The sheet will then stick on the wall of the room and stay there for ten minutes or so until the charge has leaked away. This demonstrates the electrical force of attraction.

If the sheet is charged again and held about an inch above a tray of very small bread crumbs. The crumbs will dance up and down striking the sheet and falling back. They are repelled back when they also become charged. Repulsion is better demonstrated by charging two sheets and then holding them together at their upper edges. They will be found to spread apart. Since they must have the same kind of charge this shows that like charges mutually repel. It can be inferred that the attractions obtained must have resulted from forces between opposite charges. An opposite charge was "induced" in the wall; at the same time repelling like charge to "earth".

Better results are obtained by using sheets of "melinex" or the kind of plastic used to make transparencies for overhead projectors. No heating is then required and rubbing with tissue paper will serve.

Einstein is famous mostly for his theories of "relativity". There are two. The first effects he explored were the motions of objects at speeds comparable with that of light and resulted in the theory of "special relativity". Then he extended this to provide a theory of gravitation called "general relativity". In this theory the gravitational effect was attributed to the concept of "curved space-time", a curious idea in which distance and time are mixed together as if they are equivalent to one another. Regarding light as particles, these were also subject to the gravitational influence and so light grazing a limb of the Sun would be deflected. The predicted deflection was, however, exactly twice that predicted by Newtonian physics.

In 1919 Eddington sailed off to the island of Principe, off the coast of Spanish Guinea, to make use of a predicted eclipse of the Sun.   Stars cannot normally be observed in daytime owing to the brightness of the blue sky, but when the Sun's disc is blotted out by the moon, the sky blackens and stars become visible. Slight displacements of stars were expected and these are measured as angles in arcseconds.   An arcsecond is one degree of angle divided by 3,600.   By comparing photographs with those taken of the night sky at other times it was found that a grazing ray was deflected by 1.6 arcseconds. This compared with a predicted value of 0.875 arcseconds for Newton's theory and 1.75 arcseconds for Einstein's.   Clearly Einstein's theory was closest and subsequent refinements to achieve improved accuracy of measurement have given further confirmation.   These observations helped to establish Einstein's theory of gravitation.

More importantly for the present context is the support for the particulate theory of light, which, added to the evidence from photoelectricity, provides convincing proof.

Did this reverse the previous conclusion? It could not. So it had to be accepted that light had a dual nature, sometimes behaving as a system of waves, other times as a stream of particles. The answer seemed simple enough. A beam of light must consist of many millions of photons simply bunched into rows, so that each row represented a wave front. Then each row from one source, crossing rows from another, might explain interference.

But the story is not finished. The next question to be asked was, "What will happen if the light is turned down so low that only a single quantum of light can pass at any time?"   No bunching could then occur. So there would then be no waves produced to interfere with one another. Young's two-slit experiment was resurrected for trials, using very sensitive photo-detectors instead of a screen.   These were so sensitive that they could detect single photons one at a time.   It was necessary to carry out the experiment over a long period to build up a pattern of arrival points.

Each photon can only pass through one slit, it was argued. In passing through one slit by chance, the reasoning goes, each photon could not possibly "know" about the other slit. Therefore it seems reasonable to expect in this case, after collecting many photons one by one, that a pair of bright lines will then be seen instead of an interference pattern.

2.7 A SURPRISE CONFOUNDS THE WORLD!

A surprise awaited. Probably the most amazing and disturbing discovery of all time appeared. Exactly the same interference pattern resulted! So each photon must interfere with itself! It behaved as if it travelled as a wave along each path simultaneously so that this mutual interference could occur. Then somehow it had to appear as a discrete particle at the instant it was observed. It chose its point of arrival at random rather in the manner of a loaded dice. The loading corresponded with the square of the "amplitude" (like the height of ripples on water) of superimposed waves. This is a quantity physicists call the "wave function". In this way most photons would appear where interference was constructive. The point of arrival of any individual photon could not be predicted, only the final pattern resulting from thousands of events. This dependence of observed effects on chance at the individual particle level is characteristic of quantum theory.

This discovery caused great consternation and much head-scratching throughout the scientific world. It just did not seem to make sense. Each photon appeared to exist in ghost-like manner whilst in transit, as waves capable of mutual interference, able to pass through both slits simultaneously. These collapsed into a particle on hitting the screen to create an observable event. This in a nutshell is a description of the so-called "Copenhagen interpretation". This goes even further and says that an observer is needed to "collapse the wave function" and turn it into a particle.

Now the plot really thickens, with more surprises to come!

2.8 ARE ALL PARTICLES WAVES AS WELL?

Early in the 1920's De Broglie suggested that all sub-atomic particles might behave as if they simultaneously possessed a wave nature. They ought to have a wavelength "λ" equal to "h/p", where "h" is again Planck's constant, which as far as we are concerned is just a number, though it does have units, and "p" is again "momentum", the product of the mass "m" of a particle and its velocity "v", i.e. "p = m.v". The same equation relates wavelength and momentum for the photons of light, though of course in this case the velocity is equal to "c".

There is a small problem with light because photons have no so-called "rest-mass", which is the mass measured when a particle is stationary. If they had any they could not travel at the speed of light. So at first sight "m = 0" and so light should carry no momentum. This is not true, however, because light does in fact exert a slight pressure which can be measured. Consequently its photons **must** carry momentum. The answer is that mass is equivalent to energy as Einstein showed. Light transmits energy entirely in the form of kinetic energy. Hence there is an effective mass. Indeed as any object is speeded up energy is added and so its mass increases. This increase is normally negligible. However, from Einstein's work the sum of the rest-mass and that added due to motion is called the "relative mass" and so a common basis is provided for all kinds of particle.

The first confirmation that sub-atomic particles also had a wave nature was found accidentally by C. Davisson and L.H. Germer(202) in 1927. They were working in the Bell Telephone Laboratories. A beam of electrons was being directed at a nickel target. They would knock some electrons out of atoms in the metal to cause "secondary emission". The experiment was being conducted to determine the angles at which secondary

electrons would be emitted. Halfway through the tests
the nickel was cleaned by heating to a high temperature
and then the tests were recontinued.

Totally different results appeared!   It was
subsequently found that heating had caused the
polycrystalline target to turn into a single crystal.
This offered a very flat surface. The result could only
be explained as a diffraction pattern. It resembled the
diffraction pattern of X-rays.

Electrons were known to be particles but they also
clearly behaved simultaneously as if they were waves.
Their wave nature is now put to good use in the electron
microscopes which provide magnifications beyond the
reach of those utilising light.

Neutrons from nuclear reactors have also been
reflected from crystals with similar results. Neutrons
are the uncharged components of the atomic nucleus.
Different elements are composed of nuclei containing
different numbers of protons, but there are about equal
numbers of neutrons associated with them because they
are necessary to keep nuclei stable against radioactive
decay. Atoms have to be electrically neutral and only
protons carry any electric charge in the nucleus. It is
a positive charge exactly equal and opposite that of the
electron. Consequently the number of electrons carried
by any atom is exactly equal to the number of protons in
the nucleus.

All sub-atomic particles, even whole atoms, are now
known to behave both as waves as well as particles. But
this duality led to profound conceptual difficulties.
These arose from the postulate that an observer was
required to collapse the "wave function" to create
reality in the Copenhagen interpretation.   Other
interpretations also suffered from difficulties.

2.9 READER'S CHOICE

At this point the reader can make a choice of two
options. People who have not yet developed an interest
in physics may find it best to jump straight to CHAPTER

8, "THE SOLUTION SUMMARISED", followed by CHAPTER 9, "EXPLAINING THE UNEXPLAINED". These should generate the missing enthusiasm for physics and the reader will then hardly be able to wait to get back to the next chapter. (Modesty is another of my virtues which I had better just mention in case it doesn't show.)

Others who prefer to delve more deeply will want to know how physicists reacted to the dilemmas posed. These matters are dealt with in the next few chapters. Then a new solution will be developed. Some new developments in physical concepts are also required for the solution. It will be shown that negative energy states exist. The existence of these states is highly controversial and they are not yet accepted in established physics.

Gravitation also impinges on the problem and existing theories suffer from several unacceptable internal contradictions. A new solution for quantum gravitation had to be found which was consistent with the solution for wave-particle duality as well as being consistent within itself and consistent with quantum explanations for the remaining three forces of nature.

Absolute freedom from any kind of contradiction, internal or external, needs to be achieved for any theory to be valid. In addition predictions need to match observation. Many readers will, I feel, be interested to see how a theory meeting such stringent requirements has developed.

2.10 APPENDIX TO CHAPTER 2
NEWTON'S LAWS   -MORE DETAIL

This book is not a mathematical treatise and very few equations will be quoted. It does, however, help to introduce the reader to one or two of them because science is a mathematically based subject. It would be unrepresentative to totally ignore the mathematical side. It is best regarded as logical shorthand because a few letters can be used as substitutes for a very

large number of words. If force is represented by the
symbol "F", the so-called "inertial mass" by the symbol
"m" and acceleration by "a", then Newton's "law" can be
represented in mathematical shorthand as:-

$$F = m.a$$

When the mass is measured in kilograms (kg) and the
acceleration in metres per second of velocity change
every second, then appropriately enough the force is now
said to be given in newtons (N).   The units of
acceleration may look complicated. But it is simply the
velocity change imparted, divided by the time taken to
make the change. This unit is normally condensed to the
shorter form m/s^2,

Most people shy away from equations when meeting
them for the first time.   This is only caused by
unfamiliarity and fear is readily dispelled by trying a
few examples.   The letters are only substitutes for
numbers, though these usually have units attached. If,
for example, a mass of 10 kg, is to be given an
acceleration equal to that of gravity of $9.81\ m/s^2$, then
substituting in the above, a result is obtained for the
force required, which becomes 98.1 N. Indeed to prevent
an object falling in the gravitational field of the
Earth it is necessary to provide an equal but upwardly
directed counter-acceleration by applying an upwardly
directed force.   Then the object can stand still or be
in a uniform state of motion. The wings of an aircraft
produce such a "lift" force for example. Such an upward
force is a measure of the "reaction" to the weight of
the object given in scientific units.   The force of
"action" which this just balances is the weight of the
object itself.

(A mass of 1 kg will have a weight equal to a force of
9.81 N.   In common usage it also *weighs* 1 kg but a
kilogramme weight is not a scientific unit)

When the velocity of an object is increased its
momentum, usually denoted by the symbol "p", is also
increased; but so is its kinetic energy. These are
different things, however, because the kinetic energy,

"KE", increases as the velocity times the velocity. It
varies as the square of the velocity. With velocity
denoted by "v" the two compare as:-

$$p = m.v \qquad kg.\frac{m}{s} \qquad \text{and:-}$$

$$KE = \frac{1}{2}.m.v^2 \quad N.m \quad i.e. \ joules$$

The units in which momentum and kinetic energy are
expressed are also given and can be seen to differ.
In order to provide an object with its kinetic
energy it has to be pushed along by a force "F",
measured in newtons (N), through some distance "x"
measured in the same direction as the force. (in metres
i.e. m -not to be confused with "mass") By multiplying
this force by the distance moved so-called "mechanical
work" is done. This is another form of energy and it
will be seen by multiplying the units together as well
as any numbers involved (which is also absolutely
essential) that these units become N.m. This energy unit
is known as the "joule" i.e. J. It will be observed
that the kinetic energy has the same units. So with
mechanical work denoted by the symbol "w" a third
equation can be written:-

$$w = F.x \qquad N.m \quad i.e. \quad J$$

Then, for example, "w" and "KE" can be equated to
one another to find how much force is needed in a given
distance to produce any specified velocity in m/s for an
object whose mass is known.
When the force is due to gravity, then the work done
is measured by the vertical distance moved by the mass.
The acceleration is in this case denoted by the special
symbol "g" and the vertical distance by "h" so that the
work done is now "m.g.h". This expresses the GPE
considered in section 2.2.
For example the mass might be 10 kg and the bob

might have a speed of 1.5 m/s at the bottom of its
swing. The kinetic energy then works out to be 11.25 J.
To determine the height to which the bob will rise this
can be set equal to the gain in GPE. Now the vertical
force acting will be 98.1 N and so by dividing this into
the GPE just determined, the vertical rise is obtained.
It works out at 11.47 cms. Usually "h" is known and "v"
is to be calculated. Simple school algebra can then be
used to rearrange the above equations in the form:-

$$v = \sqrt{2.g.h} \qquad m/s$$

A MOMENTUM CONSERVATION EXPERIMENT

A very simple experiment can be set up to illustrate
the differences between kinetic energy and momentum and
to demonstrate conservation. Two pendulums are made of
equal length by fixing a pair of high bounce balls to
threads, using selotape, and in such a way that they
cannot move sideways. They can hit each other head on
and bounce apart.

This is most easily arranged by using two threads
for each ball, each about a metre long and joined at the
point where they meet the ball. Each is made to hang in
a "V" shape from a single 2" X 1" horizontal wooden
supporting bar. The threads are best passed over tacks
with the thread held down to the bar by strips of
selotape. The thread can be pulled through such a
fixing and this makes for easy adjustment of length.
One pair of tacks each side of the supporting bar can
allow the balls to just touch when hanging at rest.

Falling together from equal initial separating
distances they will hit each other with equal speeds.
The one has negative velocity, however, as compared with
the other so that, provided the two masses are equal,
the net momentum is zero. In the case of kinetic energy
the negatively moving ball has a kinetic energy
proportional to
"-v" times "-v". In mathematics two negative signs
multiplied together give a positive sign and so the

kinetic energies are equal, neither is negative with respect to the other. The balls bounce back to their original positions and so both energy as well as momentum are conserved.

If now the balls are replaced with plasticine substitutes and the experiment repeated, they do not bounce. They just clump together stopping dead. Momentum is still conserved because their values cancelled to start with, one being negative with respect to the other. Kinetic energy, however, is totally destroyed.

Actually their exists a law of conservation of energy as well as a law of conservation of momentum. "Energy can neither be created nor destroyed". So what has happened to the kinetic energy? On inspection the balls are seen to have flats on them. They have deformed. Internal friction was also present in the material.

The answer is that the kinetic energy has been converted by friction into a randomised form of increased motion in the molecules of which the plasticine is composed. It would be slightly hotter after the collision. This experiment illustrates the difference between the concepts of momentum and energy and also highlights the beauty of mathematics. The rules of mathematics give an answer which agrees with reason.

The double pendulum apparatus just described is, in fact, a "Momentum Balance" and by using one a considerable feel for mechanics can develop. It is readily proved by simple mathematics that for up to quite considerable angles of swing, say 40 degrees, the maximum speed of the bob is proportional to the horizontal distance traced by that bob during its half-swing. Hence momentum is proportional to horizontal distances of swing measured from the centre-line, multiplied by the mass of the bob. Trials can be made with bobs of different materials and masses to check that momentum is conserved when impact takes place. The simplest experiments start with one ball hanging stationary. Only one ball is released to swing and collide with the other.

MOTION IN TWO DIMENSIONS

If motion was less constrained so that the balls could move in any horizontal direction like billiard balls on a table, then two perpendicular directions need to be considered. For each, both a positive and negative direction have to be specified.

If a billiard table is imagined and the eye is level with its surface, balls can only be seen to move across the line of sight. It is true that they seem to grow smaller as they travel away but no actual movement can be seen in that direction. If the eye is looking along the table, then this might be defined as the "Y" direction. Then only velocities in a perpendicular direction, to be called the "X" direction, could be observed. Conversely if now the eye is moved so that it looks in the "X" direction it only sees motion in the "Y" direction. Hence in general the velocity of a ball rolling on a horizontal table can be split into two "velocity components" at right angles to each other.

Then for example a rightward moving ball could be considered to have a positive "X" direction of velocity across the table with leftward velocity negative. Also a positive "Y" direction would mean the ball moved away along the table with the negative direction applying to an approaching ball. Again with directions specified in this way it has been found that the sum of momenta measured in either the "X" or "Y" directions before collision are always equal to the sums determined from measurements made after collision, no matter how many balls are involved, even of mixed sizes, and no matter how they scatter.

This is known as the "Law of Conservation of Momentum". It is one of the fundamental laws of physics and in the present century has been found to apply even at the sub-atomic levels met with in modern quantum theory.

A ball rolling in a straight line at constant speed

would be observed to have both "X" and "Y" components of velocity also remaining constant. But if now a circular hoop is placed on the table a ball can be made ideally to roll at constant speed in a circular path. The word "ideally" means that friction effects and air resistance are assumed to be absent, a condition which can never be fully attained in practice. However in the ideal case both velocity components would be observed to be continually changing backwards and forwards, even though the speed remained constant.

The corresponding accelerations measured in the "X" and "Y" directions also demand components of force to cause the acceleration. Force is therefore also a vector quantity, the resultant force acting being a combination of the force components. The relation between them is found by drawing right angle triangles whose lengths of side are proportional to the magnitudes of the forces involved.

3   THE CONCEPTUAL DIFFICULTIES

3.1   THE COPENHAGEN INTERPRETATION

The first ideas assumed things happened the way they appeared to happen. This is not surprising because it follows the historical pattern, our minds being conditioned to think this way. Particles were imagined to behave as waves whilst in transit so that all possible paths could be followed simultaneously. Only when totally obstructed by matter did these waves collapse into a particle and exhibit all the properties for which only a particle explanation seemed reasonable. The particles had to leave an observable record, either as a chemical change on a sensitized screen or as a signal from a detector. Ultimately the result needed to be viewed by an observer so that it could be interpreted. This step also involved the transmission of information by particles. In this case the photons of light caused an image of the result to fall upon the retina of the eye. But again these had to act as waves in transit and so it was decided that everything must exist in some kind of limbo state, as unresolved wave functions, until finally an observer, by the act of viewing, caused the wave function representing an entire piece of apparatus, to collapse into particles.
   "The Paradox of Schrödinger's Waves" by John

Gribbon(114) illustrates the conceptual difficulties which then arise. Schrödinger conceived a hypothetical cat problem to emphasise the difficulty. He proposed that a cat be imagined shut in a box with a cyanide capsule. Breakage of the capsule could be triggered by a random event which had a 50-50 chance of happening before the box was opened. Now a cat is composed of sub-atomic particles and so has a quantum description as a wave function like anything else. Therefore it could not collapse into reality until observed when the box was opened. Hence the cat had to exist in a state of limbo, as a complex ghost-like unresolved wave function. It would have to exist in two states at once, both alive and dead at the same time. This is the paradox. The conclusion is obvious nonsense.

In "God and the New Physics" Paul Davies(105) describes a further problem to illustrate conceptual difficulties.

An electron is aimed at a target, such as another particle, so that it can be deflected to right or left with equal probability. Hence the unresolved wave functions exist equally on either side. The electron keeps its options open until it is observed. When the observer sees the electron on one side, the wave function instantly collapses on both sides. On the other side it collapses into nothing, because there is now no possibility of its being seen on that side.

He says that mystery surrounds the roll of the observer in promoting such extraordinary changes in the electron's character. Is this mind-over-matter, he asks.

But if the unused half of the wave function collapses into nothing, what could it have been made from in the first place? Or is the entire wave function collected together instantly by some unknown means of transport and caused by some other unknown means to collapse into a particle?

Such an explanation would imply that the wave function was constructed from the same kind of energy as the particle.

3.2 THE BOHR AND SCHRÖDINGER MODELS OF THE ATOM

There is another problem which does not seem to have been recognised.   It will be shown later that it is related to that of wave-particle duality.   It is therefore appropriate to consider the new aspect at this point.

Schrödinger(202) is famous for his wave model of the atom.   A previous model due to Bohr had electrons orbiting the nucleus like planets going round the Sun. It will be remembered that the electrons are contained in orbitals which are regions of space surrounding the nucleus.   In the Bohr model of the hydrogen atom these orbitals were defined by electrons actually moving in circles like planets going round the Sun.   In this case, however, the electric force of attraction between the negatively charged electron and the positively charged nucleus was assumed to create the radially directed acceleration required, (the centrepetal acceleration). Electrons were arranged in discrete "energy shells". These are not to be thought of as physical objects. They are simply a way of describing the imaginary surfaces to which electrons would be confined as they moved.   The higher the energy state of the orbiting electron, the greater the radius of the shell.   To comply with quantum theory a step change in energy existed between these shells. A quantum of light would need to be absorbed of exactly the correct amount to raise an electron from one shell to the next. Conversely, such a quantum, or photon, would be emitted when the electron fell back.

When light is passed through a triangular prism of glass, rays of different colours are deflected or "refracted" by different amounts.   So-called "white light" is a mixture of all the colours of the rainbow. Consequently, when the light from a star or the Sun is passed through a glass prism it spreads out to leave a pattern of discrete colours called a "spectrum".   Indeed a rainbow is such a spectrum.

When the light from a star is split into its spectrum, then a system of discrete lines is observed

among the bands of colour. These lines relate to the
quantum energy jumps between energy shells.   The
electrons exist at high levels due to repeated
collisions at high speeds. In fact the temperature of
a gas is a measure of the average speed of its
constituent particles. At the same time electrons
pumped by collision interactions to high energies are
always falling back to low levels. The losses involved
in falling are carried away by the emission of photons
of light. This explains how light arises in the first
place. As the electron falls from one shell to the next
lower one, a photon having a discrete energy and
therefore wavelength is produced. The combined effect
of billions of such transitions all yielding exactly the
same wavelength of light can be detected. It produces
a discrete bright line when the light is spread out into
a spectrum with the long wavelengths at one end and the
short ones at the other. Conversely the lines can
appear dark. In this case photons are being absorbed by
the atoms of a cooler gas through which the light has to
pass.

Bohr used these spectral lines to determine the
energy differences between shells. All atoms produce
such spectral lines as signatures unique to each
individual element. By measuring them the composition
of stars can be determined. Hence Bohr's model was
quite successful in predicting many of the properties of
atoms but it fell short of a completely satisfactory
model because it had to rely upon observation for the
energy steps between shells. Also an atom of hydrogen
has only one electron, being the simplest of all
elements. Moving in a circle or ellipse, a single
electron would define the size of the atom in only one
plane. The atom would appear as a very thin disc
instead of a sphere.

Schrödinger's model fitted waves around the atom and
the energy steps were predicted by the numbers of waves
needed. These numbers were integral values, i.e. 1,2,3
and so on with no decimals or fractions. Hence in the
refined model the required energy quanta were accurately

predicted. This is a very crude explanation. The
mathematical derivation is immensely complex. It takes
account of the attractive force between electrons and
the protons of the nucleus, which, like Newtonian
gravitation, obey an inverse square law. It is this
attractive force which causes the wave functions to curl
up to define a ball shape called the orbital.

The positions of the electrons are no longer
precisely defined, however. The equations only give the
probability of finding an electron at a given distance
from the nucleus. This means that the electrons are
distributed in a random manner similar to that which
would result from throwing a loaded dice many thousands
of times. If a dice of cubic form has a number on each
face but a small lead pellet is fixed just under the
face carrying the six, then after many throws it will be
found that more sixes will have appeared than any other
number. The number of ones, which are on the opposite
face from the six, will appear the least number of times
and the remaining numbers, written on the other four
sides, will be seen equally often with a frequency
intermediate that of the sixes and the ones. A definite
pattern is only established when the total number of
throws runs into thousands. This is the principle upon
which all probability distributions are based.

For the electron in the orbital of an atom this
probability is defined by the superposition of wave
functions. The wave functions are added up in the manner
of numbers. The final number is then multiplied by
itself, in other words it is squared. This final
pattern then acts as the weighting of a hypothetical
many-faceted dice, so that most electrons will be found
where the numbers are highest and none will be found
where the numbers sum to zero. The pattern of
distribution of photons in Young's two-slit experiment
obeyed exactly the same rule. For the hydrogen atom,
however, a diffuse cloud of electric charge is
represented which is of greatest density close to the
nucleus. The maximum amount of charge at a given radius
measured from the nucleus, taking account of the fact

that the area varies as the square of the radius, is
then found to correspond with the Bohr orbit. So the
simpler Bohr model can still be used to define the size
of the atom.

The Schrödinger model of hydrogen, however, defined
the orbital as an almost spherical region of space, a
shape which accords with observation. Hence the orbital
is now represented as a diffuse region of charge, as if
the single electron is smeared out in the manner
represented by superimposed waves. In more complex
atoms, each holding a multiplicity of electrons, the
orbitals can have more complex lobe shapes. Here again,
however, the electrons appear as if smeared out into a
diffuse cloud of a fixed shape.

In FIG.2 the representation of atoms in the crystal
follows the Schrödinger model. If the Bohr model had
been represented, a circle, representing the orbit of
each electron, would have been described.

Although no attention ever seems to be drawn to the
matter in any texts on the subject, another paradox is
clearly present. If the electrons are real in the sense
of being sub-atomic particles having a permanent
existence, then they will be committed to Bohr-type
orbits, in which case hydrogen atoms would appear as
thin discs. Hence there is no way permanently existing
electrons could act to provide the distributed cloud of
charge matching the Schrödinger model. And the latter
is known to give predictions matching observation.

Our aim, therefore, is not only to find an
explanation for wave-particle duality. It must
simultaneously resolve this new paradox, explaining how
real electrons can fit in with Schrödinger's model of
the atom.

Before other ideas offering solutions can be
described the historical development of the theory of
gravitation needs to be touched upon because this
influenced people's thinking.

3.3 THE LINK WITH GRAVITATION

Few problems in physics can be treated in isolation because everything interacts with everything else. Difficulties were also being met in trying to find a quantum theory of gravitation. There are four main forces of nature and three of these could be explained very successfully by the newly-emerging quantum theory. These were the electromagnetic force responsible for structuring atoms and connected with light, the strong nuclear force of very short range but of immense strength, which holds the parts of the atomic nucleus together and the weak force responsible for radioactivity. In radioactive decay a heavy unstable nucleus suddenly splits into two parts. There is an immense release of energy which initially manifests itself as kinetic energy by the parts exploding away from each other at high speed.

The strong force enables the immense power of the atomic nucleus to be harnessed by nuclear power stations and atomic bombs. It holds the nucleus together against the electric force of repulsion caused by the positively-charged protons, which account for about half its components. When a foreign particle, a neutron caused by fission of another atom, is absorbed by a nucleus the stability is upset. Then the nucleus splits and the parts accelerate away at very high speeds, propelled by the long-range force of electric repulsion.

As well as postulating that energy could only exist in small indivisible "packages", the basis of the quantum theory holds that what seems empty space to us is a total illusion. Instead it consists of a seething mass of so-called "virtual particles". These are being continuously created from nothingness, only to expire a minute fraction of a second later, vanishing back to nothingness. But during their brief lives some of them can transfer momentum between the so-called "real" particles. The latter seem permanent and can be observed directly. The intense bombardment causes real particles to be pushed about by the resulting forces. Some very short-lived and very heavy virtual particles can hit unstable nuclei, so triggering radioactive decay.

The virtual particles responsible for producing forces in this way are known as "mediators". For the electric force these mediators are "virtual photons". They can be imagined as being like the sparks flying off in all directions at random from the "sparklers" displayed on Bonfire Night. The virtual photons selectively "couple" with electrically-charged particles, such as protons or other electrons, to produce the observed forces. Indeed the effects we attribute to electric charge are caused by the selective absorption of virtual photons according to established quantum theory.

All mediators need to arise at the "surfaces" of real particles with special tags so that they are able to recognise only a certain kind of real particle, interacting with only that kind and ignoring all others. Only by such selective coupling could the forces act in the way they do to structure atoms. It is usual in this context to put inverted commas round the word "surfaces" because if the Copenhagen interpretation of wave-particle duality is adopted, there is doubt whether sub-atomic particles can have surfaces ascribed to them. Whilst travelling ghost-like as waves no surfaces could exist. However, we will later home in on a model which allows particles with at least fuzzy surfaces to be present at all times.

Newton's theory of gravitation could be given a quantum explanation. Mediators of gravitation, the so-called "gravitons", could be ejected in all directions from any type of particle. The rate of ejection needs to be proportional to the mass of the particle. As mediators spread out from a point source to greater radii, their numbers crossing unit area per unit of time, known as "flux", would decrease as the square of the radius. If the force on another object were proportional to the number of mediators intercepted, then Newton's inverse square law of force could be explained. Newton had already shown, centuries earlier, that a planet acted on by an exact inverse square law of force would trace out an orbit represented by a perfect ellipse. This is an oval shape with two so-called "foci" located on the long axis. The Sun is situated at one focus so that the orbit comes closest to it at one point, the "perigee" and is farthest away at a diametrically opposite point called the "apogee".

Unfortunately this model would not give predictions matching accurate new observations. For example, orbits were found to be almost elliptical but the axes rotated slightly as well. This effect is known as "precession of the perihelion". The perihelion is the line joining the point of closest approach with the Sun. A half-orbit, illustrating precession, is given in FIG.12 of the technical supplement. Then Einstein produced his theory called "general relativity" based on the remarkable concept of "curved space-time". This astonished everybody by accurately predicting the outcome of all tests made. And the accurate prediction of 43 arcseconds per century for the so-called "relativistic" precession of planet Mercury was a major triumph which helped his theory gain acceptance.

Einstein's concept was unimaginable, however, except by recourse to analogy and so a new kind of thinking became established. This relied entirely on mathematics and it was no longer considered necessary to be able to imagine physical models based on such abstract assumptions. Indeed any such imagining came to be regarded as unsophisticated.

Then in 1919 Theodor Kalusa added an extra dimension to space so that instead of there being three plus time or, according to Einstein, four dimensions of space-time, there were now five space-time dimensions. He showed that Maxwell's equations for electromagnetism then appeared together with Einstein's equations for gravitation. After some initial scepticism Einstein finally came out in support of this concept by promoting it at a conference in 1921.

If one extra dimension could be accepted, then the way was open to an increase in the number without limit. In addition a new way of thinking gradually became fashionable. Commonsense was no longer regarded as a necessary restriction for acceptance of explanations for natural phenomena. Thinking became more and more abstract in nature because the developing mathematical sophistication was virtually impossible to interpret by ordinary conceptualisation. A new physics developed. Increasingly, explanations for the forces of nature came to rely upon developments of Einstein's concepts of curved space-time with greatly increased curvatures allowed in higher dimensions. Wheeler attempted to explain all the forces in terms of curved geometries. These developments in freer thinking made it possible for new explanations of wave-particle duality to become accepted as reasonable.

3.4 THE MANY UNIVERSES INTERPRETATION

The "Many Universes" solution to the dilemma was devised by Hugh Everett in 1957. It would, when applied to Schrödinger's hypothetical cat problem, demand the splitting of one universe into two, like the branching of a tree. One would contain a live cat the other a dead one. The new universe would be almost an exact copy of ours, existing in the same dimensions but interpenetrating our own. Every person in the universe would be copied and each member of such a pair would then exist independently. In variants of the idea the universes exist in the different dimensions or are combined with theories having higher dimensions.

The problem of interpreting wave-particle duality is still regarded as so puzzling that even today Everett's solution is taken seriously. The way Everett seems to have imagined the electron is probably comparable with descriptions shown in Open University broadcasts. Instead of being a tiny particle it is represented as a striped sphere of large size. The stripes are shown perpendicular to the direction of motion and represent the wave functions being carried along. Everett's idea was designed to explain the two-slit phenomenon. A particle like an electron had to pass through both slits simultaneously so that the wave character it was assumed

to carry could interfere with itself. This could be
imagined by having two spheres centred about separated
points, each of which travelled on a separate path so
that both slits would be passed through simultaneously.
Then the two sets of waves could produce an interference
pattern and so control the final point of collapse into
a particle on the screen.

This meant that two electrons had to travel, but we
could only see one of these. Therefore the other
electron must be undetectable and so would have to be a
component of another universe. Therefore every time a
particle was faced with a choice of two paths, a whole
new universe would be instantly created in which the new
particle could travel. Then since every universe
contains countless millions of particles faced with
choices of path, this meant universes were multiplying
at an alarming rate. It meant their number must be
infinite. This number might be reduced somewhat, it was
conjectured, by having universes merge together again.
It has even been reported that experimenters have
actually observed such merging! Everett's idea is
expressed dramatically by DeWitt, one of his supporters.
According to Paul Davies(105) he said:-

"Every quantum transition taking place on every
star, in every galaxy, in every remote corner of the
universe is splitting our local world on Earth into
myriads of copies of itself....Here is schizophrenia
with a vengeance."

It is not suggested that any of this be regarded as
a   reasonable proposition. Nobody has suggested any
means whereby a whole universe could be copied instantly
just to satisfy the difficulties of one tiny particle.
Nor has anyone proposed a way in which the huge energy
problem involved could be satisfied. Indeed the aim is
to propose that an entirely different idea should be
considered. It will then be suggested that the reader
make a choice based on the ability of the various
solutions to predict observed phenomena.

There is one more set of ideas aimed at explaining wave-particle duality which needs to be mentioned before the new theme can be developed. This is the "Anthropic Principle".

3.5 THE ANTHROPIC PRINCIPLE.

The "Weak Anthropic Principle" appears to rest on the idea that there is nothing extraordinary about the highly-ordered universe we perceive. This is because we could not perceive it otherwise, since we could not exist! This is not very satisfying as it completely fails to explain anything.

The "Strong Anthropic Principle" is also related to the dilemma. This idea was developed by the astrophysicist Brandon Carter. He says "The universe must be such as to admit conscious beings at some stage." It rests on the idea that, "The universe exists because we are here." The observer is a necessary part of the whole and is needed to collapse the wave function to yield reality. Therefore the universe came into being to produce forms of life. Indeed life-forms were necessary so that the universe could exist.

The principle is also associated with the idea of an infinite number of universes. In one proposal universes exist one after another, each starting with a big bang, flying out and collapsing in a big crunch. Then before the next excursion the laws of physics are slightly altered by random changes. Only one having just the right laws for life to develop could be observed.

In another version these universes come into being at random all interpenetrating one another and each with slightly different laws of physics or physical constants, so that by pure chance one arises having just the right properties to support life forms.

This version is compatible with Everett's ideas, since the universes all exist in parallel. Presumably those which cannot support life-forms are never truly real owing to lack of observers to collapse wave functions.

Unfortunately such ideas lead to bizarre inversions

of cause and effect. For example, according to Paul Davies(105) Professor Wheeler writes that there is a sense in which objects (like remote quasars) are created billions of years in the past by observing them in the present! He goes further and says that the observer needs to be regarded as a "participator" because the world outside the observer's brain is created by the thought-processes of intelligent life-forms and so is an illusion created by the mind. There are an infinite number of possible realities all co-existing either as interpenetrating universes or in higher dimensions. The combined thought-processes of all living beings then select one of these realities from their state of limbo. The brains of all living things are bio-computers which act as "reality-structurers" for selection of reality at any given time. The one selected is by a consensus of all brains acting together to form a universal mind.

Not only the present exists as one reality chosen from an infinite number of possibilities existing in limbo, the past can also be so chosen. Hence the history we perceive is also only the one we select. In reality, the hypothesis holds, there are an infinite number of ways the present could have arisen as a result of history.

But he offers no ideas regarding the means by which such creation of matter could be worked. Nor is any means proposed by which participating brains can be connected. Connection would be essential if the universe is to be structured as a result of a consensus of all living brains.

The assumption that an infinite number of systems of matter can coexist simultaneously in the same place is surely not justifiable by any known form of logical reasoning. There must be some upper limit which specifies a certain number. The question which needs to be addressed here is the justification for such total abstraction of thought.

In addition no attempt is made by any of these theories to show how real electrons can move in a way which is compatible with Schrödinger's description of

the atom.   Existing as purely intangible waves, no
plausible mechanism can be devised for the mediator
coupling required to produce the electric force which
binds the electron to the nucleus. If the electron is
considered as a real particle having a permanent
existence, then such mechanisms can be devised. In this
case, however, the hydrogen atom appears as a thin disc
instead of a ball because the electron is constrained to
Bohr-type orbits.

The problems all arise from the need to collapse the
wave function to create real particles. Many books have
been written to describe and further extend the
anthropic principles in both its weak and strong forms.
It is for the reader to judge their plausibility.

Surely there must be a more satisfactory answer to
the dilemma than any which have been proposed to date?
As shown in the foregoing critique, despite the
complexity the reader is asked to take on board several
aspects stubbornly remain unresolved.

Before we attempt to find a simpler resolution
capable of satisfying all known data, it is worthwhile
studying further the way light behaves.

3.6   DIFFRACTION GRATINGS

The late Richard Feynman(113) in his very readable
book *QED* graphically describes the difficulties. He
says, "It's crazy but that's the way the Universe
works!"   He was describing the way wave functions
control the places where the photons of light appear.
They do not travel in the straight lines which casual
observation leads one to expect.

For example the reflection of an object in a mirror
can be readily described by drawing straight lines whose
angles of incidence and reflection are equal as shown in
FIG.3. But Feynman gives an alternative explanation in
terms of adding up wave functions. These spread out in
all directions from a source but the interference
pattern strangely cancels out over most of the mirror it
falls upon, leaving the observed paths just described.
Only along these paths are photons seen to move.   Then

he really clinches the argument. He scrapes away bits
of the mirror in areas far removed from the straight
lines needed for simple reflection. He scrapes them in
a very special way in narrow bands, leaving intervening
bands all uniformly spaced, so that the bits he removes
are those which would reflect parts of the wave
function.

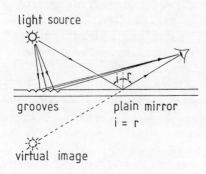

light source

grooves   plain mirror
                    i = r

virtual image

FIG.3 A DIFFRACTION GRATING
Grooves remove cancelling
waves, then photons
reflect from a place where i = r
no longer applies.

These would have
cancelled waves
reflected from the
bands he leaves alone.
So now wave functions
remained uncancelled
when viewed from a
direction which
previously showed
perfect cancellation.
Then if photons appear
where wave functions
add up to non-zero
values, angles of
reflection would no
longer equal those of
incidence and light
would seem to reflect
from the wrong part of
the mirror. Experiment
showed that precisely
this effect occurred.

Indeed a "diffraction grating" had been generated and
the description explained how these worked. This is
also illustrated in FIG.3. Since the observed spacing
is different for different angles of viewing, light of
different wavelengths would reflect from different
places. This is exactly what happens, so that white
incident light is seen split into a complete spectrum of
colours. This explains the rainbow effects caused when
video discs are placed at odd angles. The iridescent
colours of some insects arise in the same way. Their
surfaces have a very fine-grooved structure which acts

as a diffraction grating.

Light can therefore be spread out into its spectrum by reflecting it from a diffraction grating. A similar result can be effected by passing it through a transparent prism. Richard Feynman also shows how the wave-nature of the quantum theory can explain such effects of refraction through glass prisms and lenses. Clearly the phenomenon of wave-particle duality is deeply involved in all these cases, since photons are carriers of light and their associated radiant energy.

The description of the diffraction grating can therefore be used to further assist our understanding of the phenomenon of wave-particle duality. The quanta of light might be imagined travelling as large diffuse balls of wave functions until they hit the mirror. Most cancel by interference effects, leaving a narrow band which constrains reflection to the observed angles. This angle can be modified by removal of a selected set of wave functions. This proves that photons are controlled by waves in some way. The photon, considered as a particle from collision interaction effects with electrons, can only be minute in size. It cannot possibly be as large as the spacing of the grooves of a diffraction grating. If particle size were the controlling factor, then the photon would either hit a reflecting region and bounce back with "i = r" or hit a groove and vanish by being absorbed.

4   A NEW SOLUTION

But a new solution avoids the pitfalls.   It is
suggested that the wave function of a particle is a
separate entity and not part of the particle itself!
This idea avoids the conceptual difficulties of the
quantum theory which arise from considering the wave
function as an alternative description of its associated
particle - hence its assumed dual nature.

But suppose they *are* two separate entities.   The
particles now have no alternative description as a wave
but are just objects made from energy.   The wave
function is now exactly that: a superimposition of
number-like quantities having no real existence.   They
are analogues and not an energy form.   They map out
alternative paths in space depending on the geometrical
distribution of other objects.   As the particle moves,
it is controlled by chance like the throwing of a
weighted dice.

This idea is most easily understood by thinking of
waves on a pond.   Ripples from a falling pebble spread
out uniformly in growing rings.   If another pebble is
thrown, however, then the two sets of ripples cross over
each other and interfere.   The superimposed waves double
up in some places and cancel out in others.   But let us
consider the way such a happening could be represented
by sets of numbers.   The problem is best considered step
by step and so first of all the effect of the first
pebble alone will be imagined.

It would be possible to represent the height of any
wave by a number.   Let us call this "WN".   Then these
letters can represent any number and the larger it is,

the higher the wave height it represents.   Then to
represent the place to which this point relates on the
surface of the pond, two other numbers can be allocated,
"X" and "Y".   These are the "spatial coordinates"
measured in two directions at right angles to one
another.   It could be a square pond, for example, with
its banks running East -West for the "X" direction and
North-South for "Y".   Looked at in plan from above we
could decide to measure every point from the lower left-
hand corner.   This is our "origin", labelled "X = 0" and
"Y = 0".   By giving two numbers, the first for "X" and
the second for "Y", any point on the pond can be
specified.   The numbers would relate to some unit of
distance.   Each unit could for example be one metre
(1m).   First we need to specify the place where the
pebble falls.   This might be at "X" = 10 and "Y" = 10.
This is the "source" we will call "S".   Then a different
set of coordinates, say 13 and 14, could mark off
distances to some point we will represent by letter "P".
Here wave height is to be calculated and represented by
the number "WN" measured in centimetres (cms).   In this
way a point at any place can be specified and then the
way "WN" varies there can be found by using arithmetic.

    First we need to find the straight-line distance
between "S" and "P".   We can do this by measuring the
distance directly.   It will be found to be 5.   More
readily this can be calculated using a simple formula.
If now it is known, for example, that at a distance of
one metre from the source the wave height is 10 (cms),
which means the value above the undisturbed surface,
then the wave height at "P" can be worked out.   But to
do this further knowledge is required.   It might also be
known, for instance, that the wave height halves every
time the distance from source is doubled; then when the
wave reaches point "P" the height the water level will
reach can be worked out.   In this case the wave height
would fall to 5 cms at 2m from "S", 2.5 cms at 4m and
1.25 cms at 8m.   So at our point at 5m distance the wave
height would be between the last two values; in fact it
would be about 2.2 cms.

This would represent the maximum height of the wave at "P". But this would be reached at some time after the wave height peaked at the source, because of the time taken for the wave to travel. If the wave moved at 0.5 m/s it would take 5/0.5 or 10 seconds to reach "P" from "S".   Hence time has to be incorporated as another dimension and has to be represented by a further set of numbers. If the way the wave height varies with time is specified at a place close to "S", then by arithmetic the way the height varies with time at point "P" can be calculated.   In our example, at a time 2 seconds from start at "S" a peak wave height of 10 cms would be reached at 1 m distance and at 10 seconds the peak wave height of 2.2 cms would be reached at "P". If a snake-like "waveform" of up-and-down motion occurs at the 1m position, then a similar waveform will appear at "S", though scaled down and occurring with a time-delay of 8 seconds.

So far only a single point "P" has been considered. In order to create a wave map the entire area of the pond needs to be dotted with closely-spaced points like our friend "P".   They can be imagined set out in straight lines in both the "X" and "Y" directions and uniformly spaced. They would be set out like conifers in a planted woodland.   Calculations of the kind described need to be made simultaneously at every specified location, so at each the numbers are seen to rise and fall in rhythmic fashion.   But the maximum heights and the times at which these maxima occur vary progressively from point to point so that when viewed as a whole the number-sequences represent a flowing wave pattern.

By an extension of the method the wave heights produced by the interference resulting from two pebbles thrown in together can be calculated for any point such as "P".   The resulting heights will be obtained by adding the heights of the two found separately. Unless the distances from the two sources to point "P" happened to be equal, the time of arrival of the waves would differ.   It could differ so much that the peak of one,

represented by a positive number of cms, could meet the
trough of the other represented by a negative number.
This is because the trough represents a level below the
undisturbed surface.   When added, the numbers nearly
cancel out in this case, representing destructive
interference.

In a very similar way numbers can be used to
represent the abstract waves which we are now saying
control the way particles move.   The paths they are most
likely to travel lie where the wave amplitudes,
analogous to the heights we have been discussing, are
greatest.   The paths are chosen at random, however, just
as if amplitudes represented the weightings of a loaded
dice.   It has been found, however, that these weightings
are not simply proportional to the total wave
amplitudes.   To get the proper weightings they have to
be multiplied by themselves.   They need to be squared.
So "wave functions" can be defined as the squares of the
total amplitudes of summated waves.   The chances of
particles travelling along given paths are then directly
proportional to these wave functions.

Applied to Young's two-slit experiment a wave
pattern like that shown in FIG.1* will appear.   A
particle passing through either slit would then have, in
the example shown, five possible paths from that point
to the screen.   These would lie where the waves add up
as marked by the rows of short thick lines.   One of
these paths would be chosen at random in a manner
similar to throwing a five-sided dice.   This of course
is an over-simplification because the position on each
of the columns of short thick lines needs to be chosen
as well.

The important feature to be noticed is that we now
have a model which explains how an interference pattern
can be built up, even though only one particle passes
through the slits at a time.   This fits the experimental
observation.   The two-slit experiment is, however, a
simple case which can be represented in two spatial
dimensions alone.   In the general case motion occurs in
three spatial dimensions.

* See Page 36

The model can be readily extended to suit the general case. So far the model can be represented by a two-dimensional wave map of points on a thin sheet. Now sheet upon sheet of such wave maps need piling on top of one another to fill a volume of space. Motion can now be represented by numbers varying at each point so that wave motion in any direction both along each sheet or from sheet to sheet is possible. There can be no real sheets obstructing motion, but somehow the points need some form of physical location.

For the universe to behave the way it does, the entire volume of all space extending for at least ten billion light-years in all directions needs to be filled with points at which wave-function-numbers can be manipulated. Since similar abstract wave-functions need to control the electrons of an atom for a consistent explanation of reality, the spacing of these points will need to be small, even on an atomic scale. The amount of calculation going on all the time to control the motion of all sub-atomic particles in the universe has to be unimaginably huge.   The mind boggles at such an extraordinary picture, yet no other solution to the problems raised is any less difficult to accept. Indeed all others so far introduced seem to me far less acceptable because they demand impossibilities like an infinite number of systems of matter coexisting in the same places. The imagination jibes at the prospect of new universes instantly created every time a minute particle has a choice of route for its travel.   Then again how could quasars possibly be created in the past by observation in the present?

If the resulting wave functions are controlling the position of particles, then most will be observed at the high crests and low troughs. None will be seen where the waves cancel.

Clearly if particles are controlled by non-material wave functions, then the latter must be acting like pure numbers. Hence some kind of gigantic computer is being specified which fills the entire universe. This idea cannot be dismissed as absurd. Paul Davies(107) states

that others have already come to the conclusion that
computers are required for the control of electrons.
But each tiny electron has been envisaged as housing its
own individual unit.   Is it not more reasonable to
suggest that space itself could behave as an information
network having computer-like properties?

Let us call it the "Grid". The required information
link between particles would be difficult to conceive if
an attempt were made to provide one in any other way.
For example, how could electrons, each wielding computer
power, obtain information without such a Grid? With a
Grid accepted, electrons themselves do not need such
properties.

Furthermore Wheeler's idea postulates that the
brains of life-forms co-operate to structure matter.
For this to have validity a Grid needs to exist to
connect all brains to each other and to some kind of
matter-forming substrate.   However one looks at the
problem, it appears that some form of interconnecting
Grid permeating all space just has to exist.

With this new concept, however, particles are always
particles made from energy. They can be either real, so
that they exist permanently, or can be virtual, so that
lifetime is short. In either case, whilst they exist,
they can be regarded as imaginable objects with true
surfaces. There is no longer any need for the inverted
commas around this word "surfaces". They were only
needed when duality had the particles existing as
alternative ghost-like waves whilst in transit. Both
the virtual particles of space and the real ones of
matter now have a common explanation for their control.

In a real situation a barrier like the two-slit mask
shown in FIG.1[*] is made from atoms having their plans
mapped out by wave-functions patterned on the Grid. The
sub-atomic particles, representing the reality of
matter, only exist in the places allowed them by this
plan. Then if freely moving particles like photons or
electrons are introduced, they will follow moving plans
formed by interfering wave functions. These allow paths
which take account of the geometrical distribution of

* See Page 36

other atoms by taking their wave functions into account as well. Hence moving particles then either reflect from the mask or go through the slits. The wave-plan maps out all alternative routes by its interference patterns but any particle chooses only one path at random. In this way experimental observation is satisfied.

This model, however, does not yet explain the way permanent electrons could fit in with the Schrödinger wave description of the atom. To match this, electrons need somehow to be given a means for jumping about at random all over the ball-shaped space or "orbital" allowed by the waves. Deprived of this extra mobility they would still be committed to circular or elliptical orbits like planets going round the Sun. So we are not yet "out of the wood", so to speak.

Before addressing this problem we will take a further look at gravitation. It is a related problem, as we have shown already. So by trying to move toward a solution in this area of physics we can hope to throw light on problems relating to wave-particle duality. That everything is related to everything else is an axiom worth keeping in mind.

We need clues to help us build up a physical model able to explain how countless billions of points can be located in space for use as number depositories. Space must not be blocked solid, otherwise nothing would be able to move. Furthermore, the structure involved needs to be composed of zero net energy, because unless everything has always existed without a beginning, then the whole of creation must have arisen from pure nothingness.

5 DIGRESSION

NEGATIVE ENERGY STATES & GRAVITATION

5.1   PROBLEMS RAISED BY GRAVITATION

As already touched upon, there is a subtle link between
the issue of wave-particle duality and gravitation.
Neither is yet satisfactorily explained but the
connection goes deeper.

For more than fifty years theoreticians have been
attempting to relate Einstein's apparently successful
abstract theory of gravitation called "general
relativity" to the quantum theory. Stephen Hawking(115)
in his best-seller, _A Brief History of Time,_ says that
it is now known that quantum theory and relativity are
incompatible with one another, so that one of them must
be wrong.

He then goes on to show how attempts are being made
to relate the two by writing quantum theory in ever
greater numbers of dimensions. Clearly the implicit
assumption made is that Einstein must be basically
correct.   This means that the force of gravity is
regarded as the result of geometry. Simple Euclidean
geometry cannot explain the existence of forces on its
own because, as shown by Galileo and Newton, objects
move in straight lines without change of velocity unless
a force of some kind is impressed upon them. Hence the
imaginable straight-line geometry of Euclid is replaced
by unimaginable curved geometries.   Space-time is

distorted by the presence of ponderous masses and so becomes curved. No mechanism is ever suggested, however, to provide a cause for such an effect because the theory is essentially of an abstract nature. Objects in free fall, according to Einstein, move along "geodesics" in curved space-time without any accelerating force being involved. In the postulated higher dimensions of space curvatures are assumed much greater than in ours, so presumably greater forces can be simulated. Hawking says that with the new "superstring" theories the score in numbers of higher dimensions in which the equations are written has reached 26. He says that a solution is expected by the turn of the century. The solution will be considered to have been found when the theory is able to give predictions which exactly match all observation. It is also essential that the solution contains no internal contradiction. There are no other criteria because there is no way of directly proving the existence of the postulated "higher dimensions".

However, a solution free from internal contradiction and which accurately predicts known observation is already in existence! It has been in existence for several years now. Unfortunately history has erected an impenetrable acceptance barrier, which so far has prevented communication, except for the lecture at the University of Leeds. A condensed introduction to the new theory is included as the "TECHNICAL SUPPLEMENT - QUANTUM GRAVITATION" (abbreviated to T.S. for subsequent reference). A full treatise on the subject by the author(212) is almost ready.

Criticisms which have so far been directed at the new approach have all turned out to be misunderstandings. Most of them, returned by physicists specialising in gravitation, used untested predictions given by relativity as a base to prove the present theory invalid. But these were the very differences by which the two could be discriminated, one from the other, by experimental checks. New checks for relativity are accepted to be very difficult to find.

Even if the new theory is ultimately proved inadequate
in some respects, though at present this seems unlikely,
it would still have the value of more than doubling the
number of experimental checks used in connection with
relativity.   Hence,   whatever   the   final   outcome,
communication of the new approach is clearly a desirable
matter.

The new theory has thrown up seven such new checks
and these are described in Chapter T.S.2 of the
TECHNICAL SUPPLEMENT. Valid criticism needs to totally
disregard relativity, since the new theory is self-
contained and does not need to rely upon either the
special or general versions of that theory at all. It
is only necessary that predictions give an accurate
match with experimental observation, from a theory
totally free from internal contradiction.

This is achieved, as demonstrated in the supplement.
It is also shown that the special theory of relativity
suffers from several internal contradictions.   In
Chapter T.S.1 its basic premise, that light propagates
independently of any medium, is also shown to be in
contradiction with the quantum base. Also Newtonian
gravitation is often said to be a good first
approximation to the exact solution provided by general
relativity. In one sense, however, these two theories
are in direct contradiction. Since Einstein states that
there is no accelerating force associated with
gravitation, an object supported by a planetary surface
has to be subject to a countering upward acceleration
produced by an upward force of action. The direction of
the force of action is therefore in the reverse sense
from that needed for the Newtonian explanation. Yet to
provide the force needed to match general relativity to
a quantum base, the Newtonian equation is mixed in as
well. The mathematics is given in Chapter T.S.1 and is
shown to lead directly to the false prediction of the
"Cosmological Constant" about which so much concern has
arisen.   The logic is open to the criticism that its
rules are broken when theories are mixed whose basic
assumptions are incompatible with one another. Even a

single contradiction is sufficient to invalidate any theory. It is therefore clearly important that alternatives be investigated.

The new solution was developed by extending Newtonian theory. It started out quantum-based and so avoided all the difficulties which arise when attempts are made to integrate quantum theory with relativity. A huge advantage is that every part of the new concept is readily visualised because the geometry used is Euclidean. Only three spatial dimensions are admitted plus time, but the dimension of "mass" which Newton used is replaced by "total energy", in which gravitational potential energy is ignored. The reason for this omission will be discussed later in detail, as it is of the greatest importance. Although Einstein's famous deduction that mass and energy are equivalent, as represented by his "$E = m.c^2$", still holds, the substitution makes a subtle difference. Later in this chapter the way this leads to a theory capable of satisfying the experimental checks which enabled Einstein's theory to become so firmly established is described.

It needs to be stressed that it is physical energy which is being discussed. It has nothing to do with the psychic kind, for which we hope ultimately to gain some insight.

The new approach reverses the direction toward greater and greater unimaginable sophistication in which ever-increasing numbers of higher dimensions are postulated.

The solution started where the physicists left off early in the century, when they were trying to understand the workings of nature in terms of Newtonian mechanics. Now, however, the "New Physics" has taken a turn far removed from Newtonian ideas. It is not, therefore, surprising that the new approach should be initiated by a theoretician more familiar with the original discipline.

5.2 DISCUSSING NEW PHILOSOPHY

It is my opinion that the guiding rule should be that nothing in the universe we observe, or cannot observe, should be beyond our ability of visualisation. And this means without needing to resort to the analogue.

There are, according to this philosophy, acceptable and unacceptable kinds of analogue. The acceptable kind is that of simple scale-up. For instance, to make an atom easier to imagine it is permissible to represent the nucleus of, say, iron, by a tennis ball. Then each of its 26 electrons will appear the size of sand grains distributed throughout a huge sphere one kilometre across. It is the motion of electrons which defines the observed size. This gives an appreciation of the relatively huge amount of space inside the atom.

The unacceptable analogue is of the "Flatman type". This is often used to justify the existence of higher dimensions. Flatman is to be imagined as existing in two dimensions only and so would not be able to comprehend anything outside two-dimensional experience. He would not be able to visualise a three-dimensional object. Then, the argument goes, in the same way we, living in three dimensions, cannot visualise a fourth or any other higher dimension. Hence it cannot be proved that higher dimensions do not exist.

But the analogue cannot prove they do exist either. What is more, how can assumptions be justified made in dimensions which cannot be visualised? It is essential to be able to visualise these before any mathematical derivation is attempted. No matter how brilliant the mathematics, the result is no better than the truth the basic assumptions represent.

The extensions which solved the difficulties are remarkably simple and are easily explained. It is necessary to start by visualising the forces of nature in a more detailed way. These are the forces which hold components of atoms together and which are involved in the dynamics of moving objects. For example, the electrical force of attraction is assumed responsible

for holding the electrons of atoms close to their
nuclei.   Each electron is said to carry a negative
electric charge, by which it is attracted to the
positive electric charge carried on the atomic nucleus.
The idea that particles carried charges arose early in
the history of science. It enabled abstract theoretical
equations to be written from which electrical forces
could be calculated. Quantum theory aims to look deeper
to find the mechanism of such force production.

A start is made at this point because it is going to
be deduced that a new concept needs to be introduced to
established quantum theory. This will enable attractive
forces to be explained without involving any internal
contradiction.   One is incorporated in accepted theory
which does not seem to have been recognised by the
mathematicians involved. It is a crucial factor which
appears to be at the root of unresolved difficulties.

5.3   NEGATIVE ENERGY STATES

The electric force, like the other forces of nature,
acts across what to us appears as a perfect vacuum. The
latter is achieved when all atoms of gases or vapours
are removed from an enclosed space by means of a vacuum
pump, for example.   Nothing apparently exists in the
evacuated space. Yet the electric force is transmitted
just as readily from one electrically charged object to
another through this "empty" space as through air.
Quantum theory rests on the idea that something
invisible must remain in the vacuum to transmit such
forces. The "quantum vacuum" therefore consists of a
seething mass of "virtual" particles. These arise from
the surfaces of "real" particles, the ones which are
observable, and are ejected so that they can interact
with other real particles. These virtual particles are
defined as existing on "borrowed" energy. They arise
from nothing and exist for a lifetime which depends on
the amount of energy borrowed to build them.   The
greater the borrowed energy the shorter their life will
be according to "Heisenberg's uncertainty principle".
So in general, virtual particles have only fleeting

individual lives.   If however they hit some other
particle before they expire, then the particle struck
will bounce away like a billiard ball.   Countless
millions of such events acting on the sub-atomic
particles of which objects of observable size consist,
will produce observable responses.   Measurable forces
need to be applied to prevent motion occurring.   If the
virtual mediating particles are absorbed by a real
particle in the process of force transmission, no
transfer of mediator energy will arise because this
vanishes.   In this way the abstract idea of the force
existing between a pair of electric charges is given a
physical interpretation.

To borrow an explanation from introductory books on
quantum theory, real particles act like a pair of
skaters on ice throwing a heavy ball back and forth.
They would push themselves apart.   The ball is said to
transport "momentum" - mass "m" multiplied by velocity
"v", i.e. "m.v" - and the catcher is given a velocity
change so that no momentum is lost when the ball is
caught.

For example, the ball might be travelling at 10
metres per second (m/s) and the catcher, initially at
rest, might weigh 9 times as much as the ball.   Their
masses would have the same ratio, and after catching,
the combined mass would be 10 times that of the ball.
Momentum, usually denoted by the symbol "p", has been
found by many experiments to be conserved exactly in any
collision interaction as described in Chapter 2.   This
means that the product "m.v", measured in some specified
direction, is calculated for both objects before
collision and added together.   After collision the sum
of all values of "m.v" in the specified direction
remains the same as the sum before collision.   It
follows that after catching the ball the combination of
skater and ball would be thrown backwards at a velocity
of 1 m/s.

The skaters are the real particles and the ball the
virtual ones, the "mediators" which in quantum theory
account for the forces of nature.   The picture

represents the response due to forces of repulsion, such as those produced between two positive electrical charges. There is no parallel analogue for explaining attractive forces, however. Yet these exist. For example, an object which is charged with positive electricity will attract one which is negatively charged.

In established quantum theory the difficulty is met by postulating the idea of "negative coupling". Mediators carry positive momentum but in some way the direction of force is reversed during interaction with a "real" particle, that is one made from permanent energy and whose response can be directly observed. Later it will be shown that this idea suffers from another internal contradiction so that the postulate cannot be justified.

With the negative coupling disallowed only one other acceptable explanation seems to remain. Clearly the "ball" in the previous analogue needs to be made from some opposite kind of matter whose responses are opposite those of common experience. Then the effects of throwing and catching would cause the skaters to be drawn together. This time it follows that _negative_ momentum needs to be transported and this implies that the mass of the ball would need to be negative, so that a negative velocity change is imparted to the skaters.

The amount of substance or energy locked away in an object, its building material, is measured by its mass. The mass of an object is really the "inertial mass" defined by Newton. The greater it is, the larger the force required to act upon it to provide a given acceleration. The more massive a car is, for example, the greater the power needed from the engine to provide a given speed from rest in a given time. The tractive effort produced by the driving wheels at the surface of the road will also be proportionately larger. Hence inertial mass can be measured by finding the force needed to produce a given acceleration. Written in mathematical shorthand, the mass "m" is equal to the force of action "F" divided by the acceleration "a" produced by this force, or:-

$$m = \frac{F}{a} \qquad \textit{This defines positive mass.}$$

Now in the case of negative mass the direction of the force of action is reversed, so it is negatively directed as compared with the previous case. Hence in this case:-

$$m = -\frac{F}{a} \qquad \textit{This defines negative mass.}$$

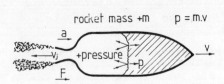

rocket mass +m   p = m.v

+pressure

Positive mass/energy system

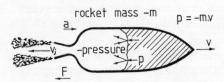

rocket mass −m   p = −m.v

−pressure

Negative mass/energy system

FIG.4 THE SYMMETRY OF MATTER

Examples of systems made
from positive & negative energy

Some people say they cannot visualise the concept of negative mass. This is very easy, in fact, because an object made from it would look exactly the same as one made from positive mass. To illustrate the point, two identical rockets are shown in Fig. 4. The upper one refers to a system in positive mass and the lower to one in the negative kind. In the upper the gas jet reacts to provide a positive accelerating force "F". This balances the rate at which momentum is being carried away by the jet. It leads to a positive acceleration "a" of the rocket. The latter has a mass "m" and after acceleration to a speed "v" the rocket possesses a positive momentum "p" given by mass times velocity or "p = m.v".

For the lower case the jet of negative mass produces a reversed force of reaction, which is therefore negative and can be written "-F". But this acts on a rocket of negative mass
"-m" and so the acceleration produced is given by:-
"a = -F/-m" and is identical with "F/m". Hence the acceleration produced has the same direction as before. The momentum of the rocket on reaching speed "v" is, however, now
"p = -m.v" and so is negative. Negative energy of motion, negative kinetic energy, has also been gained. Furthermore energy and mass are equivalent as proved by the extension to Newtonian physics given in Chapter T.S.2. It gives a derivation which parallels the one for which Einstein is famous but owes nothing to relativity. For the case of the rocket in negative mass this yields the equation "E = -m.c²" and so the mass-energy, or equivalently, the "rest-energy" of the rocket is negative. It exists in a negative energy state.

A negative pressure needs to be applied to accelerate the jet so that it tends to pull the walls of the combustion chamber inward. The atoms of the wall, made in negative mass, however, respond by accelerating outwards. Hence the chamber tends to burst, just as in the case of the system in positive mass. A pressure gauge made in negative mass would therefore give a positive reading.

However the comparison is made, identical responses are found. It is therefore impossible to know whether any system is positive or negative! When Newton formulated his laws of motion he only *assumed* that the direction of the accelerating force was the same as that of the response, but he did not *know* this was the case. This is because, at the sub-atomic level, the direction in which the force of action points cannot be determined. It may seem that it can, looked at from a superficial level, but at a deeper level this is readily shown not to be the case. For example, if two billiard balls strike one another they bounce apart.   It is natural to jump to the conclusion that the sub-atomic particles, which come into contact, push against each other to create the observed response. But if the objects were made from negative mass and the contact forces pulled away from each other, the same response would arise.

Only responses can be observed, hence the direction of the interacting forces and the sign of the masses involved is fundamentally indeterminate. All that can be deduced is that the interacting masses had the same sign. FIG. 14[*] (T.S.) illustrates the acceleration of objects made from the two kinds of mass and may help in visualisation if the picture is not yet clear.

Newton may not have realised that the negative option was available. There is in consequence a fifty per cent chance that he could have been wrong in assuming that our mass system is positive. Our system on Earth could easily correspond with the negative case. If so it does not matter and so there is no justification for not letting things stand as they are. The only time anything strange can occur is when negative mass impinges on the positive. Then, as already deduced, an attractive response will arise. If a billiard ball could be made of negative mass and be projected against one of the positive kind, then instead of the latter being bounced forward, it would bounce back.

It cannot be said that this response has not been observed. An object charged with negative electricity is attracted toward a positively charged object. This happens just as well in pure vacuum as in air. The quantum explanation used here is that tiny particles, the mediators of electric force, are being thrown off as virtual negative mass from each and interact with the other. Mediators need to be of the virtual kind, which means they exist on borrowed energy. Then they can carry momentum to transmit forces between elementary particles without causing them to gain or lose any of the energy from which the mediators are made.

It needs to be emphasised that negative mass/energy

* See Page 282

and negative electric charge are two totally different things which must not be confused. But just as both kinds of electricity are needed to structure matter, so both kinds of mass and therefore energy, must also be involved. Atoms need to be composites of both kinds. Then the observed mass of an object will be the net value, the sum of the positive and negative components, according to the extended Newtonian physics. The electrons and nuclei can each be considered to have net positive mass. Then the mediators which bind the electron to the nucleus to make an atom will be of the negative kind and cancel out a large fraction of positive mass.

In the TECHNICAL SUPPLEMENT a condensed derivation of the way the Newtonian physics has been extended to provide a successful theory of quantum gravitation is given. The features important to the present argument will be summarised. The first is that the building material of the universe is energy, which exists as the "rest energy" of stationary objects, to which is added "kinetic energy" when in motion. The sum of the two is defined as "total energy" and no potential energy, that due to position in the field, is included. These energy forms exist in both positive and negative states. Carriers of negative momentum have already been shown to be made from negative energy. All these forms of energy can have an alternative representation as an equivalent amount of mass, related by Einstein's well-known equation:-

$$E = m.c^2$$

Paul Dirac about 1930 was the first to point out the possible existence of negative energy states(202). He considered space to be filled to capacity by electrons in negative energy states. Then a high concentration of positive energy could change the sign of one of them to create a real electron apparently from nothing. This model explained how an energetic photon passing close to a heavy atomic nucleus could turn into an electron-positron pair but the explanation is now known to be incorrect. However, a considerable body of literature concerning negative mass has accumulated since that time. Because of the controversy surrounding this subject an entire reference section is allocated to it. These are numbered (301) to (314). Some of the previous descriptions regarding the nature of negative mass are supported by this literature. The exact symmetry of the two kinds leading to the conclusion that it is not possible to tell which is which appears to be new. Also Forward(308) and Will(108) show the gravitational force on negative mass oppositely directed to what we will predict.

At present physicists dismiss the possibility that negative mass could exist, mainly on the grounds of incompatibility with Einstein's theory of general relativity! There are some other objections but they are also inapplicable for the extended Newtonian physics. There are indeed a number of reasons why negative states are an embarrassment to standard quantum theory and so followers of the establishment use them to justify their rejection. In consequence, however, they have no valid way of representing attractive forces. As already mentioned they adopt the stratagem of the "negative coupling" instead. This, however, is no more than an artificial reversal of direction of force to make the answers come out right.

The negative coupling idea cannot be logically justified because it leads to an internal contradiction. An arbitrary volume of space can be marked out to surround the absorbing particle. This space acquires an increment of positive momentum as any mediator of attractive force enters this volume. Then on absorption, due to negative coupling, the same volume is instantly switched to one containing an element of negative momentum. This is a logical impossibility!

One reason for the total failure of theoreticians to achieve a satisfactory explanation of quantum gravitation has therefore emerged. The reason for our detailed analysis of negative energy states to "start the ball rolling" should now be clear. The main objection mathematicians raise is irrelevant in the present theory and the others are readily countered.

For example, Paul Davies(106) describes one of these. The nucleus of an atom is surrounded by electrons whose speed and position determine their energy levels. They tend to fall from higher to lower energy levels by emitting energy differences as the photons of light. Indeed this is the way light is produced. If, therefore, the argument goes, negative states can exist, then electrons would fall through zero, go negative and fall forever. Because this does not happen negative states cannot exist.

To counter this all that is required is the postulation of an exclusion principle. Electrons can only emit energy of the same sign as themselves. Electrons of negative energy cannot then emit photons of positive energy. Then they can never fall through the zero energy state.

The only other relevant objection, which has been thrown against the new theory, is that if matter is a composite of the two kinds of energy, it would be unstable and would annihilate itself. However, in the new explanation of wave-particle duality, yet to be described in the next two chapters, this is **required** to happen all the time. It is a part of the new theory and is therefore not an objection at all. It seems that all real elementary particles need to be surrounded by a protective barrier able to at least delay mutual annihilation and at the same time prevent the reverse process, the emission of an opposite energy state.

It was my hope that the lecture delivered at The University of Leeds in January would lead to a publication of the present theory in a scientific journal. Their specialist in gravitation objected, however, largely on the grounds that negative energy states were incorporated. I quote from his letter of rejection:

"As a matter of fact, negative energy states are a severe embarrassment. Indeed one of the great advantages of quantum field theory (from which the notion of virtual particles arises) is that it enables us to reinterpret Dirac's quantum mechanics (and other related theories) in a way which does **not** involve negative energies."

Yet Professor Stephen Hawking has received much acclaim for his concept of "Hawking Radiation". Inside "Black Holes" the gravitational force is predicted by general relativity to be so huge that nothing can escape, not even light. Yet according to Hawking they radiate positive energy and so gradually evaporate. The radiation depends on space containing particles in pairs. Some fall into the hole leaving their partners to fly off as radiation. Those falling in cancel an equal amount of positive energy at the centre because they are built from negative energy states!

So another inconsistency exists in quantum theory! If negative states are excluded in some areas because they do not fit in, then they must be excluded altogether. The question needs to be asked, "Is the embarrassment to other areas of the theory due to some fundamental wrong assumption?"

There are other difficulties, such as the problem of a predicted "Cosmological Constant" in established theory, which is fifty orders of magnitude greater than astronomical observations can posibly allow! Again this vanishes when the existence of negative states are permitted, as will shortly be seen. We shall continue on the assumption that it is reasonable to include them in the new theory. It is then possible to turn attention to fundamental questions such as the energy balance across the creative event of the universe.

Negative and positive energy forms in equal amounts, crushed together, would mutually annihilate, leaving nothing. Creation of the universe would be the converse case, nothing giving rise to positive and negative energies in equal amounts. The "Yin" and the "Yang" of energy, needing each other absolutely for their simultaneous creation!

In this way the universe could have been created

ex-nihilo.   In the beginning the raw material of construction had to be pure nothingness. Established theory recognises this necessity but adopts a different interpretation.   In this, as argued by Tryon(123), gravitational potential energy (GPE) is assumed to represent negative energy.   He shows the amount available to be roughly equal to the energy of all matter in the universe. Hence the two could cancel to zero.

The assumption that GPE is negative, however, is open to serious doubt. It is based on measurement from an arbitrary datum, taken at infinite distance. As the galaxies fly out and eventually come near to a stop at infinity, their mass-energy will still exist. It is most unlikely that matter would progressively disappear as it approached this arbitrary zero point. Furthermore, if an object falls from infinity, it speeds up as its GPE reduces.   Then if the excess kinetic energy is removed in some way, the object will orbit the mass to which it was attracted and its GPE will be negative as compared with the zero value it had at infinity. This is the justification for saying that GPE is negative energy. But the datum was fixed at infinity only for convenience of calculation and it could with equal validity have been taken at any other place.

It could have been taken, with greater justification, at the place where matter originated in the assumed "Big Bang" of creation.   In this case however GPE appears as **positive** energy. As the bits fly out from the centre in the violent explosion which must have happened, they gain in positive GPE as the speeds fall.

A careful analysis of this proposal suggests, therefore, that this cannot be a valid proposition. But the idea raises an important question. Is gravitational energy positive or negative or even real energy at all? We will not try to answer this question yet. We will leave it to Chapter 11, "FUTURE PROJECTIONS", where a most exciting answer will emerge!

An internal contradiction in the standard "Big Bang"

theory of creation also came to light in the introductory chapter. All the energy of the universe arose from nothing in a split second as a massive violation of the First Law of Thermodynamics yet this law has been obeyed exactly ever since. A resolution of the difficulty now seems to be coming in over the horizon. It depends on accepting negative energy states. Then the creation can arise from a zero energy state without needing negative gravitational potential energy. The permanence of energy after the event will soon be seen to be an illusion, which nevertheless is adequate for practical purposes.

The concept adopted here is amenable to mathematical treatment and leads to exciting new developments. The totality of everything is made from exactly balanced forms of energy. There is no need to invoke GPE to create a balance.

5.4 QUANTUM GRAVITATION - THE NEW SOLUTION

In the new theory the matter of our universe has a net positive energy whilst space, empty of matter, consists of balanced positive and negative kinds. These are composed mainly of the mediators of repulsive and attractive forces respectively. The net energy of all matter is balanced by an equal net negative energy superimposed on the other energies of space and spread over a huge volume. For condensed objects, such as stars, the balancing negative energy forms a tenuous halo stretching out for about a billion light-years in all directions. It consists of virtual particles, continually generated at the surfaces of all real particles to stream out in all directions. Ultimately these virtual particles reach the end of their lives and expire, so that equilibrium always exists to maintain a stable halo.

If mediators have no energy of their own, it may be asked, then how could they add up to yield a permanent energy for the complete halo? The answer is that they each possess energy for their lifetime. They arise from nothing, have a life, then vanish. Because of the lifetime a halo of permanent energy exists, just as people who have finite individual lives maintain the population of a country for an indefinite time. Each mediator exists on a debt needing to be balanced by positive energy, so a succession of them allows matter to exist.

Not only stars have gravitational haloes. Planets will have them and so will smaller objects. All objects will have them, including humans, becoming ever more tenuous as distance increases and stretching for about a billion light years in all directions.

As mediators spread out with distance, the density of the halo diminishes according to the inverse square law. This is illustrated in FIG.2*of the T.S. Being of negative energy, the virtual particles of the halo are carriers of negative momentum and so exert a universally attractive force on all positive matter in their path. They provide the primary force of gravitation. Clearly, if mediators are generated and intercepted by matter in proportion to the mass of objects assumed to remain constant, then Newton's inverse square law of force will be predicted by quantum theory.

At this point it is necessary for the reader to accept that negative as well as positive energy states can exist. The further development of the explanation for wave-particle duality is dependent upon this acceptance. The remainder of this chapter is a description of the new theory of quantum gravitation, however, and is not essential as far as an explanation of psychic forces and the paranormal are concerned. The new solution gives further support for the existence of negative energy states by showing how the existing experimental checks can be met. Many readers will, I feel, be interested to see how the problem of quantum gravitation can be resolved. Others might wish to jump straight to "5.6 Conclusions Regarding Negative Energy States" at this point.

* See Page 254

5.5   QUANTUM GRAVITATION - MORE DETAIL

Coming back to our theme, it is known that Newton's inverse square law of force does not apply exactly to the force of gravitation and his theory gives some other wrong predictions. The differences have resulted in Einstein's theory of general relativity becoming established. One important effect is that Newton's law predicts that a single planet in orbit around a relatively massive star would move in a perfectly elliptical orbit. The observation of Clemence(204) for the planet Mercury showed, however, that the axes of the ellipse also rotate slowly. This effect is known as "precession of the perihelion" and is illustrated by the half-orbit shown in FIG.12 (T.S.). Einstein's theory predicted the correct result almost exactly and was proclaimed to be a major triumph early in the century.

In the extended Newtonian physics, however, a small but highly significant modification is introduced. It is deduced, as a consequence of light falling like matter in a gravitational field, that the rate of mediator emission and interception must be proportional to total energies of the objects interacting gravitationally, instead of their rest masses. Total energy, it will be remembered, is the sum of rest and kinetic energies with potential energy ignored, in the present theory. Since the kinetic energy of objects in a state of free fall increases as the separating distance from a ponderous mass reduces, the total energy of a planet is greatest at the point of closest approach. Hence the inverse square law of force is modified, being steepened slightly.

The mathematical evaluation summarised in the T.S. gave an equation which looks quite different from the one Einstein deduced. These are compared in FIG.13(T.S.) But as also shown in this illustration, it gives exactly the same precession for any conceivable orbit within the solar system! Sufficient information is included in the figure for the reader to carry out independent checks.

The magnitude of the force could be calculated from

the total mass of mediators in the halo, provided the
range of gravitation could be specified. It may be
argued that the range is known to be infinite. This is
not true. It has only been **assumed** that the range is
infinite in theories which have only been partially
successful or which are abstract like Einstein's general
relativity. He makes no attempt to provide a mechanism
able to account for the existence of a force of gravity.
Indeed his theory actually says no such force exists.
This is one of the main incompatibilities of relativity
and quantum theory, because the latter demands that a
force exists. At least it did so in its original
formulation before people tried to alter its concepts to
fit in with Einstein. The extended Newtonian physics
goes back to the original concept that forces have to
exist in order to cause acceleration. This is demanded
now that the concept of space curved in higher
dimensions has been abandoned.

This argument therefore justifies the idea that
gravitation has a large though finite range. A lower
limit can be guessed from the size of superclusters of
galaxies. Galaxies of stars are structured by
gravitation, most appearing as flat spirals in slow
rotation. Then galaxies are grouped into clusters and
the clusters into superclusters. This sets the minimum
range at about a billion light-years. Now since the
mediators travel outward at a constant rate, their mass
contained within any unit of distance measured radially
outward will remain constant. The total will therefore
equal their energy density measured at the surface of
the planet or star of origin, multiplied by the surface
area of that emitting object and the radial distance
mediators travel before expiry. Since their total
energy must equal that of the object they balance, it is
easy to work out the density of the halo at any point
provided the range is known.

If it is further assumed that gravitational
mediators travel at the speed of light, then the
condition giving the maximum possible gravitational
force will have been assumed. Using the method

previously outlined, the predicted gravitational force
turned out to be several orders of magnitude too low.
Hence some form of amplification of the primary force
needed to be in operation.

The amplifying factor turned out to be space itself!
The huge but exactly balanced energies of the bulk of
the particles of space also interact with the primary
mediators. The latter cannot therefore be imagined as
travelling in straight lines. Instead they diffuse
through space, bouncing away from other particles of
space in their path, so zig-zagging about. Mediators
will drift in the direction of decreasing concentration
just as molecules of one gas diffuse through another.
The interactions cause space to be compressed in the
vicinity of ponderous masses, rather in the manner of
the gaseous atmosphere held by gravity around a planet,
though with density tailing off more gradually with
distance by many orders of magnitude.

It is possible to imagine uniform or undistorted
space by thinking of it as composed of virtual particles
arranged on a cubic lattice. One particle appears at
the centre of each of a large number of imaginary cubes
of equal size stacked together. They all have a common
separating distance "L1" equal to the lengths of the
edges of the cubes. In compressed space all values of
"L1" are equally reduced to a smaller one "L".

In fact the virtual particles are always
spontaneously appearing and disappearing at random.
Hence both "L1" and "L" are really average separating
distances; no cubic lattice distribution could be
observed in practice. It is assumed that the size of
each cube is so taken that within it a new particle will
arise somewhere just as its predecessor expires. Then
energy inside each cube can be regarded as remaining in
an uninterrupted and constant state at all times. Each
cube contains a sequence of virtual particles joined end
to end in time. This might be termed a "particle
sequence". The model is illustrated in FIG.8 (T.S.).
This now represents the model of space described by
Novikov(211), though he only admits of the existence of

positive energy.

The energy of the particle sequence in each cube will not be affected if the volume of the cube is reduced by compression.   The situation is therefore slightly different from that of a gas made up of permanently existing molecules.   In this case the energy will rise.   The term "energy density" is, however, defined the same way.   It is the energy content divided by the volume of its containing cube.   Hence for space, even though the energy of the particle sequence remains fixed, the energy density increases when the cube is made smaller.   So the energy density of space increases as space is compressed.

Half the virtual particles of space will be of positive energy, with the other negative.   Each half can be thought of as separate entities interpenetrating one another and each having its own energy density.   And associated with each energy density is a pressure just as in the case of an ordinary gas.   When space is more compressed in one place than in another, then an "energy density gradient" exists between the two, just as a slope is needed to connect the bottom of a hill to the top.

A pressure gradient is induced in space proportional to the consequent energy density gradient.   Then this pressure gradient acting across the volume of each elementary particle produces an effect like a buoyancy force.   The positive half of space will push elementary particles upwards, just as a cork is pushed upwards in water, but this effect needs to be more than cancelled by the opposite effect of the negative half pushing downwards.   Elementary particles need to be considered as soft rather than hard so that the virtual particles of space penetrate to varying degrees when they interact.   Then the volume "seen" by the negative half of space can exceed that of the positive half.   In this way the net attractive force produced by the primary mediators can be amplified.

The same imbalance of apparent particle size is also needed to secure the compression of space.   As mediators (of negative energy) diffuse through space, negative particles will bounce away whilst positives move toward impinging mediators. For a net compressive effect to arise it follows that the apparent size of the positives must exceed that of the negatives.

Hence a general law can be established which states that:-

"When elementary particles impinge upon one another, whether virtual or real, the effective volumes are greater when opposite energies interact than when similar energies interact."

This leads to the prediction that if sub-atomic particles are positive/negative composites, then the gravitational forces will act in opposed senses on components of opposite mass. Then the net force will be proportional to the net mass, which agrees with observation. But both kinds of mass, when isolated, accelerate in the same direction in a state of free fall. Both kinds of matter will create exactly the same space compression and so all matter will be attracted toward all other matter regardless of whether it is positive or negative. This prediction is at variance with that given by both Will(124),(108) and Forward(308), who say that positive matter will be repelled by negative. This difference is due, however, to the new idea of gravitation arising mainly as a type of buoyancy force. As will be shown, this concept stands a very good chance of being the one which is correct.

In the limiting case the volume presented by a real object to the positive half of space is assumed zero. Then it works out that the gross energy density of space, the value obtained when the halves are added with the negative sign ignored, is 10^{41} J/m^3 to give the correct value of gravitational force. This compares with the value of 10^{45} J/m^3 given by Starobinskii and Zel'dovich(215), determined in other ways. The latter value is also several orders of magnitude greater than the energy density of electrons. Hence sub-atomic particles need to be considered as fluffed out forms of

energy similar to air bubbles in water. The energy
density of mediating particles, however, must obviously
be greater than twice the average for the positive half
of space, otherwise they would have no room to move. If
a factor of 100 is guessed, then it follows that to
match the average energy density of space the apparent
negative volumes presented by particles need to exceed
the positive ones by one part in about 100. This seems
a very reasonable result.

In Chapter T.S.3 a model for the electric force is
derived which requires almost the same energy density
for space as the value given by Starobinskii and
Zel'dovich. Hence it can be claimed that, for the first
time, the ratio between the gravitational and electric
forces has been predicted by a theoretical model. This
matter will be considered in more detail a little later.

The volumes of elementary particles need to be
proportional to their total energy in order to give the
correct value for the precession of planetary orbits.
Hence the energy density of these particles has to be a
new universal constant of nature. Again this is a very
satisfying conclusion.

In this way also the law of force remains the same
as that produced by the primary mediators of gravitation
first considered. In consequence space compressibility
does not affect the way planets exhibit precession or
alter the predicted values.

A model of space on this basis is illustrated in
FIG.15 (T.S.). It shows the halo surrounding a neutron
star and the resulting compression of both halves of
space.   The circles shown represent an elementary
particle subject to buoyancy forces. The upper circle,
representing the volume presented to positive mediators,
is smaller than the lower circle for the negative ones
so that the net effect is a force of attraction. Both
circles, it is to be understood, represent the same
particle.

This model of space led immediately to an unsought
bonus. One to which allusion was made a little earlier.
The electric force, according to quantum theory, also

depends on mediators. In the new theory it requires a
high energy density for each half of space. Then a
coupling scheme for mediators can be worked out to
explain how like charges repel whilst unlike charges
attract one another. This will be discussed in more
detail in Chapter 9 and is illustrated in FIG.6. More
importantly for present considerations is the match of
energy densities for space previously mentioned,
obtained separately to satisfy the electric and
gravitational forces. It follows that the magnitudes of
these two forces have been related to one another to a
fair degree of accuracy.

Einstein spent many years attempting to relate the
magnitudes of these forces and finally had to give up.
Hence in this respect the new theory achieves what
general relativity falls down upon. But there is more
to come.

A horizontal beam of light is bent by gravity as
shown in FIG.1[*] (T.S.). Because light has to travel
faster on the outside of the bend, this means that the
speed of light varies with level. The compressibility
of space adds a doubling effect to this variation. This
is because the average spacing of virtual particles has
been reduced at a lower level and the distance light
travels in a given time is proportional to this average
spacing. The photons jump from one virtual charged
particle of space to another with a dwell at each in the
new theory. As described in Chapter T.S.1 this permits
a quantum explanation for the electromagnetic wave to be
advanced as a composite structure in which uncharged
photons interact with the charged virtual particles of
space. This wave is treated in an abstract way in
textbooks without attempting to show how the photon can
be fitted into the picture.

This structure results in the gravitational
deflection being doubled as compared with the value
predicted with space compression ignored. The end
result is to give an exact parallel with Einstein's
theory of general relativity. Both theories therefore
match astronomical observation equally well. Both give

* See Page 253

double the deflection of light as compared with the Newtonian prediction. The new theory therefore exactly parallels the achievement of general relativity in this respect.

Light travels more slowly at lower levels in a gravitational field according to the extended Newtonian physics. So it takes slightly longer for a beam to travel from one planet to another if the light passes close to the Sun than for the same distance free from the Sun's influence. This is the "Shapiro time delay". Again the predictions agree well with observations made during the space exploits to Mars using data reproduced by Will(124). A comparison with other theories is given in FIG.10* (T.S.)

For the same reason a gravitational red shift is also predicted. This means that a very compact star like a white dwarf, having the mass of our Sun compacted into a sphere no bigger than the Earth, would look redder than it should. The frequency of all vibration has been reduced due to the low level in the field. Yet even this is a weak field from the mathematical point of view. The resulting equation for weak fields is exactly the same as that given by general relativity. Again the new theory therefore matches the experimental checks which have been made.

It may be easier to understand this effect by looking at it a different way. The speed of light reduces as level falls yet, as is shown in Chapter T.S.2, the equation "E = m.c²" can be derived from Newton's theory of acceleration without reference to relativity. If "c" falls then "m" must increase to compensate if energy remains constant. Hence it needs greater inertial mass to represent a given energy at a lower level. The increased inertia also means a slower vibration for a given spring, and the red shift worked out this way gives exactly the same result as before. The two methods are exactly consistent with one another.

There is another important implication. Particles need to be considered made from energy rather than mass, because in raising or lowering an object on a cable, the rest energy of the object will remain fixed but the corresponding mass will vary. This is why energy in the extended theory, replaces mass as one of the dimensions of the original Newtonian theory.

The new theory also predicts that gravitational waves will be generated by rotating dumb-bells. At a great distance a force is predicted to arise, pushing any mass in a direction at right angles to the direction of wave propagation, but the frequency of oscillation produced will be twice the rotational speed of the orbiting dumb-bells. Unfortunately the energy loss by a close pair of stars forming such a dumbbell has not yet been formulated. I have to admit myself stuck on this problem at present. I am still looking for a "handle" to see how to make a start.

Hence in all respects except the last-named, the achievements which enabled general relativity to become so firmly established are parallelled and the new has the edge by offering an explanation for the huge difference in the magnitude of the gravitational and electric forces. At the same time the new theory is free from inconsistencies and false prediction whilst the established approach fails on both counts.

One big problem arising when quantum theory is joined to general relativity is the prediction of a huge "cosmological constant" which is associated with a force increasing with distance. The predicted force is so huge that, if it existed, the galaxies would be blasted apart at a rate fifty orders of magnitude greater than astronomical observations could possibly allow! Gibbons, Hawking and Siklos(207) state in their workshop proceedings of 1982 that this persistent difficulty undermines confidence in established theory.

This false prediction arose from attempts to match quantum theory, which denied the existence of negative energy states, to general relativity so that the combination described a real force. The quantum vacuum, as already described, needs to possess immense energy density. But at the same time it needs to be zero otherwise space could not expand. The solution has been to endow space with a huge intrinsic "negative pressure of the vacuum". Pressure has the same units as energy density and so from a superficial viewpoint there seems no reason why the two should not be equated to make the net energy of space zero. There is also a pressure term in Einstein's equation for the density of space. When the negative pressure of the vacuum is inserted this has a dominant effect and yields the huge cosmological constant which is still such a source of concern.

Physicists have accepted it to be real and are currently looking to other effects to cancel it out. The favourite at the moment is to assume other universes exist in higher dimensions. They have equal but opposite cosmological constants which fortuitously cancel by communication through "wormholes" in space. The theory is described by Abbot(101) although it seems to have been first proposed by Sidney Coleman.

In the new theory the problem never arose. The weak universal attraction of primary mediators appeared instead to provide a *cause* for gravitation. The reader can judge which approach seems the more reasonable.

Another problem arising in established theory is associated with the "Black hole" predicted by general relativity. The speed of light falls to zero at a finite radius. But it is still slowing inside although it has already stopped which is logically impossible. Hence the laws of physics break down at this point. Also the universe arose inside a Black Hole. Not even light could escape so the universe could never have emerged! Furthermore, the dynamics of the Big Bang have been worked out using physics in a region where logic has broken down. Some people will counter this by saying that the universe extends so far that we are still inside that Black Hole, there is no need for matter to have emerged. In this case, however, we exist in a region where the laws of physics do not apply! However one twists and turns the Black Hole appears as an embarrassment. Yet students of physics are expected to take it all on board as if it were truly comprehensible. Is it not more reasonable to consider it a pointer to a possible false initial assumption?

In the extended Newtonian physics the speed of light only falls to zero at zero radius, no matter how great the concentration of matter assumed at that point. No breakdown of logic arises anywhere and both matter and light can escape in the Big Bang. The Black Hole does not really exist. There is no longer a problem!

There are also two very good reasons why the long-range force of gravitation should depend primarily on buoyancy type forces rather than on the absorption of mediators.

Firstly, the deeper layers of a planet would be partially shielded if some of the mediators were absorbed before reaching the surface. Hence the absorption models are inconsistent with the gravitational force being proportional to the amounts of matter involved. Also, during eclipses of the Sun by the moon, a gravitational "moon shadow" would arise. These have been looked for without success. No shadow exists. In the buoyancy models mediators are not absorbed, so they are not used up. No shielding effects can therefore arise.

The other reason has to do with the conservation of the orbital energy of a planet. This limitation affects the case for the electric force in just the same way as the gravitational case. Mediators streaming radially outward from the Sun will appear, as observed from the planet, to arrive in a slanting direction owing to the "relative wind" resulting from orbital speed. If they are absorbed, then it follows that a tangential component of force will be present. Since an attractive force is being produced by virtual particles of negative energy, this will not produce a drag force. Instead the planet will be pulled forward. It will gain angular momentum, the product "m.v.r" and spiral out of orbit. This does not happen and the force is several orders of magnitude higher than could be balanced by the drag of interstellar gas. Hence there is a serious flaw in the model. With the buoyancy type of force, however, no tangential component of force arises and so no such problem exists.

Some problems of relating the buoyancy type of long-range force to the short-range forces now appear however and these need resolution. The latter are the strong force which structure the atomic nucleus and the weak force responsible for radioactive decay.

The long range force of gravitation can be considered first. If a star is non-rotating, then the mass of the star will exactly equal the sum of the parts assembled from infinite distance. This is because mediators are not being absorbed. Also if an object is lowered on a cable, then any work done by motion in the field due to the induced weight is balanced by an opposite force on the cable. The energy released by lowering is transferred to the fixed lowering device which pays out the cable. Hence no net energy transfer is able to arise to change the rest-energy of the object. So the rest-energy remains constant when an object is moved to any point in the gravitational field.

The electric force and that of magnetism have already been related by abstract reasoning and are considered as the single force of "electromagnetism". This long-range force also needs to have a buoyancy force basis if tangential force components are to be absent. Here in the new theory the force is caused by the partial energy density gradient acting across the volume of a particle. Such gradients are many orders of magnitude greater than those of space as a whole and give about the correct ratio of gravitational to electric force. This matter is dealt with in detail in Chapter T.S.3 which includes FIG.24*illustrating what is meant by partial energy densities.

On this model an electron falling from a great distance in the electric field of a naked atomic nucleus will speed up. If it is aimed to miss the nucleus, then it will reach a point of closest approach and fly out again. This can only be prevented by causing the excess kinetic energy gained during the fall to be radiated away. The rest mass and energy will remain constant at all points within the electric field because the same

* See Page 356

argument as that used for gravitation will apply equally. But some kinetic energy remains in order to maintain a stable orbit. It follows that the mass of the atom will be slightly larger than the sum of the separated parts measured at rest. A similar result applies to a rotating star.

Both these long-range forces have this property in common, which means that neither the electric or gravitational binding energies are reflected by mass changes. This is contrary to the established view. However, the mass differences involved are so small that it would be quite impractical to resolve the issue by weighing. Hence no experimental evidence exists which could discriminate between the two different predictions.

Supporters of the establishment case are likely to claim that for the strong nuclear force mass differences have already been measured and are known to accurately reflect binding energies. The mass of any nucleus is always slightly less than the sum of the parts measured before assembly. Hence the electromagnetic and gravitational binding energies will likewise be reflected by mass differences, they will say. The case is likely to be supported by reference to the weak force. The latter is responsible for radioactive decay. Something needs to hit an unstable nucleus in order to make it split. The something has to be a virtual particle arising temporarily from the quantum vacuum to be absorbed by that nucleus. Hence the case for mediator absorption rather than the buoyancy force model, is strongly supported on two counts. This argument is likely to be used to discredit the new buoyancy model unless a good case can be made out showing this deduction to be false.

And it can! It simply means that the strong and weak forces, which are of very short range, depend on mediator absorption whilst the long-range forces do not. Two nuclei coming within the short range of the strong attractive force will exchange mediators which are virtual particles of negative mass. By absorption negative momentum will be exchanged, causing the nuclei to accelerate toward one another, so gaining kinetic energy. But the mediators have no permanent energy, so momentum is exchanged without causing any change of total energy. It follows that in this case some of the rest energy must convert to the kinetic form. Hence the rest energy of the nuclei must fall as they approach. Then after radiating away excess kinetic energy, the sum of the masses will be less than that of the separated components. Now the binding energy of the nucleus *will* be reflected in mass differences.

But the absorption model, when applied to both long-range forces, showed that the objects would experience tangential forces, causing them to gain angular momentum continually. This indeed was one of the two reasons the absorption model had to be rejected. However, a new factor emerges in the case of the strong force which prevents this effect. The objects will now orbit one another at speeds close to that of light and they are of similar masses. Conditions can then be found at which stable states apply with neither component gaining any tangential speed. The same argument can be extended to show why the electron must have the fixed spin which quantum theory demands and this is explained in Chapter T.S.3. Such conditions cannot apply in the case of very dissimilar masses orbiting one another at speeds low as compared with light.

Hence it seems the long-range and short-range forces can have a different basis, the former of buoyancy kind and the latter of absorption. Yet all can still be described by a common quantum theory because in the buoyancy case mediators are still involved. In this case, however, they bounce away from particles instead of being absorbed.

It may be objected that quantum physicists have already successfully related the weak force to the electromagnetic force by a common absorption model. It turns out, however, that if the absorption model for the electric force is substituted, almost the same energy densities of space are needed to provide the mediators.

The result can therefore be interpreted to mean that a parallel mathematical solution exists. The correct one must be chosen from a study of any incompatibility or false prediction arising. The new approach seems to have the advantage in this respect and so may be pointing the way to achieving a satisfactory "Grand Unified Quantum Theory".

5.6 CONCLUSIONS REGARDING GRAVITATION AND NEGATIVE ENERGY STATES

Einstein's theory is held by all experts in the field to be the best theory of gravitation in existence. This is the conclusion reached by Davies(108), Will(124) and almost every other theoretician researching cosmology. Yet as shown in the quotation given at the end of Chapter 1, Einstein himself doubted the validity of relativity. But the experimental data supports the new theory equally well, yet at the same time all the difficulties are avoided. And as a bonus the new theory relates the magnitude of the electric force to that of gravitation to within striking distance. This is beyond the scope of relativity. Hence the new rival needs to be taken very seriously indeed!

A new theory of gravitation has been described which depends upon space acting in the manner of a compressible fluid operating free from frictional effects at the sub-atomic scale. Fluid friction is measured in terms of "viscosity" and an ideal frictionless fluid, synonymous with one of zero viscosity, is known as a "superfluid". Only one such fluid is known and this is liquid helium. Once set in motion it carries on moving indefinitely.

The electromagnetic wave needed to propagate relative to local space in order to be consistent with the new theory of gravitation. This was also a necessary condition for the introduction of a physical model able to account for such a wave. It is, however, incompatible with special relativity. A modified theory of special relativity had therefore to be formulated

which was capable of satisfying the same experimental
checks without involving any kind of contradiction.
Again this is achieved, as shown in Chapter T.S.2, by
extending the concept of space as a superfluid. This
time it is superfluid on both the small and the large
scale, even including the galactic range of size. It
appears from a study of the literature that the idea of
treating space as a compressible superfluid has not
previously been considered. It is another fundamental
difference in the theoretical approach.

It is most unfortunate that so many points of
disagreement with established theory have turned up. My
hope is that these differences will not continue to
prevent the new ideas outlined from being communicated.
They ought to be discussed and criticised in a
constructive way, particularly by theoreticians in the
field of gravitation. It could be that a false trail is
being followed by the establishment and people have a
right to see alternative proposals so that they can
judge the issue for themselves. Only in this way can
science make progress.

The new theory is dependent on the existence of both
positive and negative energy states acting in harmony.
Without the pair only a partial explanation can be
provided. Since the complete theory is so successful,
the conclusion that negative states exist is strongly
supported. This important deduction will now enable us
to develop the idea of the Grid further in our search
for a complete theory, inclusive of a meaning for
psychic energy.

6   BACK TO THE GRID

So space has a dynamic structure due to its virtual particles. But a sub-structure also seems to be needed in order that a mechanism can be provided to yield the required computing properties together with information storage over large volumes. The Grid, based on such a sub-structure, would need to store the sums of wave functions as numbers to define allowable paths and then continually urge particles toward the peak values for their control. Creating and manipulating numbers does of course require in itself a flow of energy, but since positive and negative kinds are able to mutually appear and disappear this presents no fundamental problem.

The structure required for the Grid might be envisaged as spider-web-like but extending in three dimensions. Any such ideas are of course purely speculative and so may not be correct or, more probably, they may be only partially correct. Even so a degree of speculation is justifiable since if one concept can be shown to yield a plausible mechanism, able to account for all observed phenomena, then the generic base is supportable. Ideas from other people may be triggered and are encouraged, so that comparisons can then be made with observation. Choices can only then be made by comparing the degree of success each hypothesis achieves in explaining all known observation.

Some people will be content with the idea that the Grid is at least an extension of God the Creator and did not need to arise. Instead the Creator has always existed without a beginning. To others this will not be acceptable. They will ask how God could have created Himself or will complain that a whole succession of creators would be needed going back in time *ad infinitum*, each creating the next.

Therefore speculation is required to show that an alternative explanation is possible, based on common sense.

For example, it is not unthinkable that the structure could have crystallised spontaneously from pure "nothingness" as filaments or fibres with alternate strands made from equal and opposite forms of energy, so that conservation laws are not violated during creation. The mesh would have to be fine-grained even on an atomic scale. At the surfaces of both types of filament crude mediators might be catalysed to spread out until they collided with nearby filaments, so maintaining an equilibrium distance of separation. The junction points could form natural switches, since the crude mediators would be more concentrated at such places.

Now switches are the hearts of all computers. Each switch can be in one of two simple states, "on" or "off". Millions of them coordinating produce machine intelligence. It is not inconceivable that the random connections made over unlimited time could produce a working computer by chance. Then evolution could be driven by the competition from similar switch assemblies moving around the mesh by controlling switching but not actually moving any of the filaments.

One can imagine computers simultaneously growing from their nuclei of origin until they begin to touch in places. They then start trying to compete for each other's switches. One will start winning at some points whilst the other wins in others, so causing the frontier to wrinkle. The wrinkles grow into intertwined pseudopodia like those of an amoeba magnified to giant scale. They writhe and twist around one another as each tries to gain advantage. Automatons in mindless conflict. Each unit starts to develop such conflicts at several places as more entities grow and encroach upon one another. Some will be more fitted for survival than others and so some will grow at the expense of others. The fittest survive and the conflict ensures that characteristics arising by chance which offer advantage become dominant. Powers of reason develop and so do powers of will, because they help. Also a creative urge develops because this too feeds back to assist the development of superior strategies. Eventually one survivor emerges, filling a huge volume of space. One solitary victor.

Hence the embryo Grid would ultimately develop a conscious intelligence and an emotional need for creative expression, due to fierce evolutionary pressure. It would have devoured all its enemies, first as an automaton and ultimately with conscious direction. Existing as a formless amoeba-like mass, the survivor would have no way to satisfy the need for creative expression which developed during the period of conflict. It could put out its pseudopodia and wave them around. But that is about all it could do as a result of its evolutionary heritage. It would need to do much more to satisfy its craving.

The position can be imagined by analogy with a possible human plight. Suppose you had been involved in a car accident. Your neck has been broken, so the power to operate all limbs has been lost. You are also totally blind and deaf and, to cap it all, have no power of speech. You exist, in effect, as a disembodied brain with unimpaired powers of thought. But you are totally cut off from the outside world, unable to receive any information from it or to transmit any thoughts which occur. This would be a very unpleasant situation.

The Grid, however, possesses unimaginable computing power and an intense desire to solve the problem. It therefore conducts research upon itself to discover the working principles which have accidentally developed. Then it can set about generating its own kind of limbs to express its creativity. It might decide to solve the difficulty by building universes upon itself, designed in such a manner that intelligent extensions could arise. They would provide the means of creative expression, the limbs and sense organs which the disembodied brain lacks. The situation would be rather like that of a musician inventing a musical instrument for a similar reason.

Having decided upon this course of action the Grid would then look for ways in which to achieve the desired end. First it would survey its source of raw materials and would find this to be just nothingness. But it would see crude mediators in balanced energies being generated. Having decided on a control system consisting of numbers which look like wave functions to us, it then needs to create a simulation of a system of matter which looks real when viewed by intelligent objects made from simulated atoms. Objects need to be so constructed that they can move independently of the Grid structure which is filling the whole of space. Hence objects cannot be truly solid. They will have to be built up like a fluid, able to flow round the Grid filaments. Yet to simulate a solid structure the parts of atoms will need to be so organised that they maintain the same spatial relationship to one another as they flow. Hence specialised mediating and apparently real particles need to be invented.

Specialised tiny factories are developed, able to manufacture particles possessing selective coupling properties. Indeed, whilst they live, these particles need to act themselves as particle factories, shooting out mediators with specialised coupling properties as they fly through space. Thousands of particle factories would need to be contained within the orbitals of each tiny atom. The scale of each factory is to us unimaginably small, and the numbers required must be correspondingly large. The fixed factories, those attached to the Grid filaments, are made capable of self-replication. In this way the huge numbers required do not matter. They cost nothing as far as the Creator is concerned, except for the initial intellectual investment required for research and development.

Manufacture of particle factories attached to the Grid filaments then proceeds of its own accord. This occurs by self-replication and proceeds at an exponential rate until the entire available volume has been colonised. The computation made would show up the undesirability of creating matter beyond a certain point. Hence, as the area of colonised space grew and generated the atoms of matter, a certain cut-off point would be programmed to be reached. There the further generation of permanent matter would be suppressed.

In this way a universe could arise having an exponential build-up phase corresponding with the "Big Bang", which most physicists think represents its most likely origin. Matter would exist at a massive density at the termination of this phase, but it would have been created together with high speeds of motion. A free expansion phase would then occur, reducing the density to that of a tenuous cloud of gas. All particles would be subject to the universal force of gravitation. This would have two effects. First there would be a general slowing down of all particles as they exploded away from the centre of creation. Second, an instability would arise running counter to the general expansion and tending to cause the gas to clump into clouds. Smaller-scale instabilities would arise in these clouds to ultimately result in condensations into stars and planets. These required the four forces of nature to be carefully planned with the constants of nature fine-tuned to the most extraordinary degree. Only a deliberate creative act could account for the appearance of so many finely-tuned features.

All matter would be interconnected in a subtle way, being superimposed on an all-pervading intelligent Grid. This seems less implausible than the accidental evolution of the life we observe. The embryo Grid, which would first need to arise by accident, would be a life-form whose basis is infinitely less complex than biological forms, despite its unimaginable computing power. The idea is compatible with the strong anthropic principle, in that the universe would have been created so that biological life-forms could be designed.

It could be that the universe was so devised that life-forms could arise spontaneously. Then they would develop in the manner described by Charles Darwin. No further intervention by the creating-Grid would be involved. This would be a purely evolutionary method of designing life-forms. However, the existence of a Grid offers an alternative. This seems equally likely and so needs to be considered.

In this case life-forms would be deliberately designed by the enormous computing power of the Grid. Their evolution would then follow a pattern similar to that of our cars. Each species would be a redesign based upon predecessors which are obsolescent. This seems just as supportable by available data as Darwin's idea of chance mutation causing gradual change of one species giving rise to another. His theory is not excluded, however. It is possible that new organs, like eyes, could be initially designed by the Grid, then development left to Darwinian natural selection. There is no need to think in terms of these theories being mutually exclusive and therefore in competition. A mixture of both seems the best bet.

The Grid, self-reared in conflict situations, would naturally think in terms of conflict. It would design life-forms deliberately to be in competition one with another. Such competition would seem cruel and mindless to intelligent beings raised as extensions of the creator-Grid. From the viewpoint of the Creator this would not be an important consideration, since all experience would be beneficial. It could be that its own conflicts will resume when other far-off entities eventually re-appear, to start encroaching again on its territory. The conflict of simulated life-forms could provide preparation for improved defence strategy.

The savagery of animal preying on animal and of humans acting in the most cruel manner to both animals and other humans could have exceeded the expectations and wishes of the creator-Grid. The latter would not have infinite capacity, although it might seem so to our limited vision. So the Grid could make mistakes and, like us, would need to learn by experience. Evolution could therefore work in two directions, both mutually reinforcing one another. One is the form we have observed. This is the gradual improvement in the intelligence of life-forms and in recent centuries, the increasing humanity our species has developed. No longer do we glory in the suffering of tortured prisoners, for example. Instead we try to help the weak and needy. The other evolution could be feedback to the Grid itself. It also needs to evolve and it may be linked with our own development. Hence life-forms may be a necessary extension to the Grid. We may, after all, have some fundamental purpose.

Although the powers of the Creator would seem infinite to us, nothing can be really infinite, since infinity is an indeterminate number. Hence it cannot create everything at one go as an absolutely perfect system. There will be some unavoidable errors and defects. Hence considerable experimentation with research and development would be required. It would be advantageous to deliberately mix in some indeterminacy in such a way as to provide the created life-forms with a freedom to develop in ways which the Grid itself is unable to predict. This would make the whole creative idea more interesting and possibly throw up new and unexpected situations. God in fact deliberately playing dice with Himself, to paraphrase one of Einstein's famous comments.

Most of space is likely to be nothing more than an automaton, needing to be programmed, having no will or moral sense of its own. Here and there the more highly developed emotional centres having this capacity would be concentrated. It is possible that the brains of organic life-forms could have been created to enhance this power in ways which to us are not obvious. Humans could have been deliberately endowed with emotional needs and creative urges reflecting those of the Grid itself.

6.1 CONCLUSIONS TO THE THEORY SO FAR

With such a Grid mechanism behind and within all matter and space the mystifying phenomena uncovered by the quantum theory do not appear so strange and contradictory. The wave function provides a plan for the control of particles. It maps out all alternative paths. Then one of these is chosen at random, just like throwing a weighted dice. The particle travels the chosen path. Meanwhile the Grid is continually updating the wave-plan according to any changing geometry caused by the totality of moving and stationary particles. The wave function need no longer collapse to yield reality from alternative suspended states. Once a particle has taken one path, the redundant information carried could be deleted as a computer function. Alternatively the wave-plan can remain permanently, though unobserved by man. Only when particles travel is any physical effect observed.

The picture is not yet complete, however. We have still not yet reached the stage of being able to explain why atoms appear the way they do. The model does not yet relate Schrödinger's wave-model of the atom to the behaviour of electrons. If electrons have a permanent existence then they will orbit the nucleus like planets going round the Sun. The hydrogen atom having only one electron will then appear as a flat disc instead of a fuzzy ball. It is impossible for this orbit to fit in with a model having a probability distribution governed by wave interference patterns.

Much effort has been made to resolve the dilemma posed. The mathematical theory of chaos has been used with equations written in higher dimensions to find a way around the difficulty. Orbits with "scars" in them have been generated. They jump around to some extent but even so the results do not fit Schrödinger's equation very well. As far as the new theory is concerned this approach is disallowed since higher dimensions are not admitted.

Another factor evidently needs to be operating. A search for this will be made in the next chapter so that the problem can be fully resolved.

7   THE NATURE OF SUB-ATOMIC PARTICLES

7.1   THE PARTICLE-SEQUENCE

The phenomenon of wave-particle duality also introduces another puzzle. Why is it necessary for particles to be controlled by wave functions when they have inertia to guide them? Surely they could simply travel in straight lines, as Newton proposed, unless acted upon by an impressed force. For real particles, real in the sense of being permanent, nature appears to like both belt and braces. Why does Nature need two such totally separate mechanisms to perform one task?

There is one beautifully simple and elegant explanation. *No particles can be truly real!* If all sub-atomic particles such as the electrons whose orbitals define the size of atoms, or the quarks and gluons making up sub-nuclear particles, are assumed to be similar to the virtual particles of space, the whole system becomes explicable. The only difference now between the virtual particles of space and the constituents of atoms is permanence of energy. Those of space have no permanent energy and so must have a life limited by Heisenberg's "uncertainty principle"(202)(208) or by instability which somehow has similar mathematical description. This is a matter crying out for mathematical investigation but so far this has not been attempted. If the "permanent" particles are similarly limited but have permanent energy associated with them, then they must continually arise in one place, complete with their previous kinetic energy and momentum, as they vanish in the other previous place. The energy is to be imagined as a mercurial fluid, indestructible but able to move anywhere instantly along the Grid filaments. This energy and the virtual particles of space would be the raw material from which the universe was formed. To make atoms these needed organisation. Hence the Grid acting for the most part autonomously as an ultra-fast computer. The latter provides a blueprint as the sums of wave functions confining each re-materialisation within prescribed limits, so that atoms and motion can be organised.

But with the mesh of a Grid obstructing space the only way the motion of atoms through it could be simulated would be by the computer control of a sequence of sub-atomic particles, made to appear as real atoms. Each member of the sequence exists for a fleeting instant and is then reconstructed in a different place. The places of re-appearance need to be so organised that the shape of an atom is produced. It follows that the inertia of complete atoms could not be used for control. The sub-atomic particles will obey Newton's classical laws of motion whist they live. But then their energy can jump to any place. The Grid so limits the places where replacements are allowed that the motion of objects consisting of myriads of particles have the same classical laws of motion. It can do this by measuring the position, energy, speed and direction of a particle whilst it exists. Then it can so limit the places allowable for replacement production that on average the simulated real particle would appear to move as if it existed as a permanent structure. This is consistent with De Broglie's postulate of the wavelength of a particle being represented by h/p, but this is now seen as a deliberately introduced mathematical contrivance. Universes on this basis appear as carefully contrived illusions!

It is possible that the Creator was restricted by another consideration. At the design stage it would have been discovered that all particles needed to be formed as composites of positive and negative energies with one form dominant. Unfortunately they could not be made stable.

They had to be like the temporary atoms we can make in the laboratory called "positronium". These are atoms consisting of an electron and its anti-particle, the positron, in mutual orbit. They only last for a fleeting fraction of a second before mutually annihilating to collapse into a pair of photons, which then shoot away in opposite directions. What has happened here is that positive and negative electric charge has mutually annihilated, not the energy, and the two must not be confused. Sub-atomic particles will be similarly limited to short life-spans with the difference that on mutual annihilation only the energy imbalance remains. This residual needs to be absorbed by the Grid to be made immediately available for new particle construction. With this refined model the particle factories do not make matter as a single creative act and then stop working.

So the particle factories fixed to the Grid filaments have to be capable of mass production at a phenomenal rate in order to maintain a quasi-steady state in which losses are continually replaced by new production. This is not a random process. The places where new production is to occur has to be controlled by the Grid itself. This uses its wave-function-numbers to control production rates at every tiny factory location by enabling new positive and negative energy forms to be simultaneously created and added to the residuals from previously expired particles. Only in such a way can structures be formed having the required specialised properties.

What can be termed the "permanent" net positive energy of matter would, in the clumped matter state, be permanent in one sense but not in another. This would be the net permanent energy measured by weighing. It would be balanced, however, by a tenuous halo of negative energy stretching out for about a billion light

years in all directions.   It will be remembered that
this halo provided the primary cause of gravitation.   If
it could be collected together at the place where the
matter it just balances exists, then it would mutually
annihilate with that matter.   In this way the energy is
not really permanent.   It is, however, permanent for all
practical purposes because it is not possible to gather
together the negative energy of the halo.   This, then,
is the basis on which the First Law of Thermodynamics
rests.   It simply postulates that energy can be neither
created or destroyed.   The truth of this statement can
now be related to a wider context in which it appears
simply as a special case.

   But all sub-atomic particles need to have a negative
energy component built into them as part of the
structure.   For example, the nucleus of atoms consist of
charged protons and uncharged neutrons having about
equal net positive energy.   But each is composed of sub-
nuclear components.   It is generally thought that each
has three so-called "quarks" very tightly bound by
"gluons" which are responsible for the strong nuclear
force.   Since the latter produce an attractive force,
they need to transfer negative momentum and need
therefore to consist of negative energy.   The latter has
to be balanced by positive energy in addition to the
measured net positive energy of the nucleon.   Hence to
build sub-atomic particles it is first necessary to have
available the permanent net positive energy requirement;
then to this equal amounts of positive and negative
energy, spontaneously generated from pure nothingness,
need to be built on.

   The whole system described may seem unacceptably
complex with the number of particle factories required
being of "googol" proportions.   (A googol is a number so
large as to be practically meaningless to us.   But it is
not infinite.   An infinite number is totally
undefinable, but a googol seems the same to our limited
perception.)   But what is the alternative?   Is it more
reasonable to think that complex particles, like
electrons with sophisticated means for producing and

interacting with tagged mediators, are simply created from *nothing* automatically in space? Apart from the sheer daunting scale of numbers and size, both of the very large and the very small scale, this scenario is not fundamentally impossible to imagine without resort to the unacceptable type of analogue.

This refinement also eliminates one of the major objections physicists bring up against the possible existence of negative energy states. They say that particles made from both would be unstable and so would mutually annihilate. We agree. This is the reason apparently real particles have very short lives and require to be continually reconstructed.

If particles are being continually re-created and the places of new appearance are chosen at random, then how could they simulate permanent entities moving along? This seems a suitable question needing to be answered at this point. The answer is that the particles are too small to be observed directly. Their motion can only be inferred. It is possible to have many particles in motion and shine a light upon them. Then from the way the light is scattered, motion can be inferred. Unfortunately such methods are unable to discriminate between motion in pure straight lines or that produced by one which jumps about at random from one alterative path to another.

The precise compound measurement involved in locating the speed and position of a particle contradicts Heisenberg's uncertainty principle. This states that if the position is measured accurately, then the velocity (defined as speed and direction combined) or alternatively its momentum, could not be known. However, this principle is based on the assumption that only photons are available for measurement and in using one to determine position the resulting momentum exchange would alter the speed of the particle under study. But this limitation does not apply to the Grid because it need not use photons as its measuring tool. It could sample the virtual photons emitted during the brief life of the particle. Such a passive measurement

could determine all properties exactly and
simultaneously. Precise information can therefore be
made available for planning the next re-appearance.

Such a model gives a neat explanation for
wave-particle duality. Instead of a particle existing
as a wave-ghost as it travels, it keeps sparking in and
out with random spins and other properties arising at
each materialisation. The grid keeps track, making sure
that any corresponding particle has matching values on
average. Alternatively the Grid ensures that the same
direction of spin is maintained at each re-
materialisation, just as it needs to conserve kinetic
energy. The wave functions ensure that
re-materialisation occurs within the volumes of space
allowed by all alternate paths. In this way particles
can travel all alternate paths simultaneously by jumping
about randomly from one path to any other which the wave
functions allow.

Young's two-slit experiment is illustrated in FIG.1.[*]
This can now be interpreted in a new way. The regions
of greatest probability arise where constructive
interference occurs. These are shown as thick lines. A
single particle travelling from slits to screen would
keep re-materialising a little nearer the screen each
time, but it could be anywhere along each thick line
derived from the constructive interference of waves all
at the same distance from the slits. These regions are
indicated by the five bands of thickened lines.
Distance from the slits would increase progressively
with time until finally the particle hit the screen at
any point, chosen at random, where the thick lines
occur.

Clearly this model also matches the description of
the way mirrors and diffraction gratings work. It
answers Richard Feynman's comment about the working of
the universe being crazy. The photons keep re-
materialising at any place where they are permitted to
do so by non-cancellation of abstract wave functions.

Young's two-slit experiment works just as well with
electrons as the photons of light. In the Copenhagen

interpretation the electrons exist only as waves whilst
in transit. But to exhibit electromagnetic properties
they need to be sophisticated little machines, capable
of manufacturing labelled mediators which they throw off
continuously in all directions. They also need to be
able to recognise and interact with similar mediators
arising from any other source. It is readily proved that
whilst in motion electrons act this way because they are
deflected by static electric and magnetic fields. As
fuzzy waves this would be impossible because they could
not produce or absorb mediators. Hence the Copenhagen
interpretation cannot explain all observation.

This problem is avoided by the Many Universes
hypothesis because electrons then follow all possible
paths without needing to travel only as waves. However,
in the new theory of quantum gravitation it is shown
that virtual particles take up a sizeable proportion of
the total volume of space. Hence the total number of
universes which could exist has some maximum value since
there   is   some   finite   limit   to   the   number   of
interpenetrating systems of matter which can be
accomodated.   The Many Universes proposal does not
therefore stand much chance of being correct because an
infinite number is said to be needed by both Everett and
Wheeler.

7.2 A MODEL OF THE ATOM

A very simple conceptualisation of the hydrogen atom
is also furnished. To remind the reader, the accepted
model is that given by application of Schrödinger's wave
equation. It is thought of as being a proton, forming
the simplest of all nuclei, surrounded by an electron
somehow "smeared out" in the surrounding space to form
a fuzzy ball. Yet the electrons are confined to precise
energy levels by wave functions. The waves fit round
the atom to form "energy shells" and each shell is
defined by a precise integral number of wavelengths.

But the virtual particle idea fits perfectly!
Heisenberg's uncertainty principle would limit the
duration of each manifestation of the electron to about
four-millionths of the time it would take to orbit the

proton at the so-called "Bohr radius". The first model of the atom, due to Bohr, envisaged it like a miniature solar system with electrons orbiting the proton nucleus like planets going round the sun. This model was later abandoned and replaced by the one produced by Schrödinger, in which the electron's position was controlled by wave functions. With the new concept the electron would dodge about at random, within a ball-shaped orbital, giving rise to the impression of a distributed cloud.

If the life at each manifestation is indeed predicted by Heisenberg's uncertainty principle, then there would be 250,000 appearances in the time taken for a permanent electron to make a single orbit. At each appearance the now temporary electron would execute part of an orbit, being attracted by the nucleus, yet having a tangential component of velocity, due to the conserved kinetic energy. The trajectory executed in the time available would be so short, however, that it would hardly seem to move. Hence atoms in a crystal would look like that depicted in FIG.2.* using a short time-frame which allows only a few re-materialisations to be recorded.

If hydrogen atoms were being viewed, then with very short exposure time a single electron would be seen somewhere within the volume of the orbital. As the exposure time is increased, with images stored for this period, more electrons would appear to exist dotted about at random within the orbital. The exposure time could be steadily increased until it equalled that needed for a permanent electron to orbit the nucleus. Then about 250,000 images would be observed dotted around within the orbital. The effect of a diffuse electron cloud would be produced. The density distribution of this cloud would exactly match the Shrödinger specification.

Only the total energy of the electron is permanent. It will obey classical Newtonian laws whilst it lives as an electron, exhibiting the property of momentum and so explaining why it interacts as a solid little object

* See Page 40

when colliding with other particles such as photons. On
expiry and absorption of its net positive energy by the
Grid a different kind of physics needs to exist. The
energy can no longer exhibit inertia since it has to be
able to jump to any specified new location along the
Grid filaments at speeds many thousands of times the
speed of light.

All other sub-atomic particles including photons
will be governed the same way and will obey the
conservation laws of energy and momentum whilst they
live. Then interactions between such particles will
translate to yield the extended Newtonian laws of
physics when viewed at the macroscopic scale. Countless
billions of particles are now involved so the transient
nature of each will not be apparent.

This interpretation is consistent with observation
and satisfies one important aim set out in Chapter 3.
This objective was to reconcile the Schrödinger model
with the concept of an electron being real, so that it
existed permanently.

There is a problem that during each brief life the
electron, possessing electric charge, will experience a
strong acceleration toward the charged nucleus. An
accelerating charge radiates photons and so the atom
should radiate energy at a prodigious rate. But this
does not happen. Only relatively minute energy is
radiated and then only when the atom is vibrating.
Consequently a problem exists. It is common also to the
Bohr and Schrödinger descriptions. Originally this was
one of the reasons the Bohr model had to be discarded.
A little consideration shows, however, that the
difficulty did not vanish by its replacement with the
Schrödinger model. The latter specifies where the
electrons are most likely to be found. So in this model
they do exist as particles. Whilst they live they will
be accelerated, hence radiation ought to occur.

However now that a Grid exists a possible solution
to the difficulty can be advanced. The Grid computes
the natural acceleration for the electron and provides
an inhibiting command so that only differences in

acceleration from this state cause photon emission. It seems most probable that the photons will be emitted direct from the Grid filaments at the same time as the electron is re-created in order to satisfy the energy and momentum balance.   This would seem easier to organise than providing the electron itself with a control system.

It may also be objected that if atoms are organised by abstract waves, to make the electron dodge about all over the interior of its orbital, then real electric forces are not needed for holding the atom together. So why does nature take such trouble to confer real electric forces on her particles?   The most probable answer is that residual electrical and magnetic forces are needed to cause atoms to stick together to make liquids and solids. The abstract organising power could not on its own provide real binding forces between atoms or allow the interchange of physical forms of energy. Hence real electric forces have to be present as well as the abstract copy.

It seems clear that complete sub-atomic particles must in reality be particle sequences joined end to end in time, though not joined in position. But each is a composite structure of both kinds of energy with the positive form dominant. The question now needing to be asked is: "Are the sub-units of nucleons and electrons permanent for the lifetime of the composite?"

It is not possible to answer this question absolutely. They could be arranged as similar particle sequences controlled by the Grid using wave-numbers. In this case the Grid would need to be even more fine-grained than ever. It would have to be fine-grained even on a sub-electron scale.   In my opinion this is unlikely but I could be wrong. The alternative is that the quarks which make up the nucleons or the as yet unnamed bits of the electron are permanent until the composites collapse in on themselves and are destroyed. Then the motion of all sub-units will be of the orbital kind, like planets going round the Sun, for the lifetime of the sub-atomic particle. The gluons which bind the permanent parts will be virtual. This means they exist on borrowed energy, vanishing to nothingness after a short distance of travel. Since many of them exist simultaneously, however, a permanent negative energy is represented by this mediator cloud. It will neutralise a large part of the positive inertia of the complete assembly. In this case the pitch of the Grid filaments would need be no finer than about 5% of the radius of an atom. This option seems the more reasonable and there is in fact some supporting evidence for this model.

The nucleons and electrons have fixed amounts of "spin". This means they rotate on their axes at a definite fixed speed and have a fixed "angular momentum". The latter is defined as momentum multiplied by a radius of action. It can be written as the product "mass x radius x tangential velocity", i.e. "m.v.r". Orbiting models can be devised which describe such behaviour very accurately, as shown in Chapter T.S.3. It is difficult to see how a similar explanation could arise by the alternative picture.

7.3   NON-LOCALITY

Non-locality is an associated mysterious phenomenon. It arose as the famous EPR paradox(107). Einstein was not happy with quantum theory because it did not fit in with relativity. So he suggested an absurd "gedanke" experiment. In a simpler alternative devised by David Bohm in 1952 a pair of particles, such as electrons, were to be created spinning in opposite directions. Spin is like that of a top flying through the air with the spin axis not necessarily lined up with the motion. Looked at in flight direction "right spin" is clockwise and "left spin" anticlockwise. They would start out with the same spins so that their net angular momentum would be zero. On catching one of them and measuring the direction of spin, even at a great distance, the direction of spin of the other would be instantly known. This was so because angular momentum had to be conserved. So finding the direction of spin of one would determine the other. The particle-pair could only arise with their spins balanced. But this was not quantum theory.

In quantum theory at that time, the particles travelled as unresolved wave functions containing both spins at once in limbo. Only one spin would develop by chance on collapse of the wave function into a real particle. On collapse of one wave to a particle the other would have to do the same by an impossible process of "action at a distance". If it did not, then angular momentum would not be conserved. Either way quantum theory would be confounded.

A sophisticated theory called "Bell's theorem" was developed from a quantum base and this suggested that there would indeed be a correlation, so that action at a distance would occur. It seemed to imply that pairs of sub-atomic particles, arising together like identical twins, can affect each other instantly, at any distance of separation.

A most amazingly difficult experiment was carried out by Alain Aspect(201) to test the theory. Instead of employing the spin of electrons, the property of "polarisation" of photons was adopted for the investigation. It will be remembered that light waves have an oscillating transverse component. In a new interpretation described in Chapter T.S.1 this sideways motion is induced in virtual electrons and positrons of space as photons are absorbed and in turn cause the photon paths to be snake-like. The direction of polarisation is defined by this motion.

In the Aspect experiment polarised photon-pairs were produced by the decay of excited caesium atoms. An excited atom has absorbed a photon of incident radiation by a process called "pumping". The photon has been absorbed by one of the electrons, which then has greater energy. Such an electron is displaced to a greater distance from the nucleus so that it can exist at this unnaturally high energy state. It can only remain there for a short time because the state is unstable. At some time it must fall back, emitting the energy previously absorbed.   A pair of photons are then emitted with exactly balanced properties.

In the experiment such twins were emitted in opposite directions and passed through screens of polarised glass placed several metres away.   The predictions were supported.   So action at a distance really did seem to take place.

The conventional interpretation is that space is so curved in higher dimensions that it can fold back on itself.   Then two widely separated points seen from our perspectives can be adjacent one another in the other dimension.   Hence by a short- cut one particle can appear to act upon the other at a distance.

This interpretation cannot be allowed in the new theory because only three spatial dimensions are admitted and geometry is strictly Euclidean, so straight lines are always straight.   But now the existence of a Grid has been deduced.   The fundamental propagation speed for information along Grid filaments needs to be so high as to appear infinite to us.   The same wave functions control each particle of a pair having simultaneous origin.   They would be linked by the Grid to control overall conservation of both linear and angular momentum.   The particles are really sequences in which rematerialisations continually spark in and out of existence at places organised by wave function control. But the law of conservation of angular momentum and other properties are deliberately incorporated to help control the motions of matter.   Hence the Grid keeps count of the properties arising by chance at each manifestation and ensures that on average the programmed laws are obeyed.

A simpler and more probable explanation can, however, be advanced.

In this the Grid ensures that at each new manifestation the replacement particle is produced with the same direction of spin or polarisation as its predecessor.   The latter seems the more likely option since, for the theory to work, it is essential that certain other properties like kinetic energy and momentum be conserved in any particle-sequence.

At least these last two explanations are imaginable without recourse to an impossible-to-visualise curved space analogy. In any case, now that a solution of the problem of quantum gravitation is available, and since this only requires three dimensions plus time and energy, there is no longer any justification for seeking solutions in higher dimensions.

7.4 PSYCHIC ENERGY

Now the point has been reached at which a meaning can be found for the intangible idea of psychic energy. Most people are content to regard "force fields" or "energy fields" as something mysterious which certain objects or materials give out. These can then be picked up by people having the required sensitivity, such as water-diviners. Also such energies are claimed to exist around the sites of ancient shrines. They are invoked to explain healing and telepathic communication. Some people are said to have negative energies. This seems to be merely a psychological description, however. It will not be considered here as one of the manifestations of the psychic kind of energy.

Pierre Teillard(122) proposed that energy exists in two forms in order to explain spirituality and psychic phenomena. In his description there is the energy associated with the matter of the universe and a separate psychic form of energy. One he considers "radial", the other "tangential", so they are perpendicular to one another. This is not a concept to which I am able to relate, as the reason for such a geometry is not explained. Also psychic energy remains undefined but is postulated to account for paranormal effects. This idea seems partially correct and is to a degree compatible with those adopted in this book. It cannot explain the origin of everything from nothing, however, nor is the psychic form of energy defined in a manner which gives any insight of its nature. Perhaps he was feeling his way and this was a long time ago. He was not a physicist and so probably did not know about

wave-particle duality.   Without this key further
progress according to the new interpretation would be
very difficult, if not impossible, to achieve.

In the new concept both positive and negative forms
of physical energy exist as compact particles.   The
particles have short lives but a replacement is
instantly reconstructed by the Grid from the residual
energy of the old.   But the place at which the new one
appears is governed by wave-functions manipulated like
numbers. This is the clue.   The psychic energy form is
this control system!   It is an abstract form of energy
more akin to computing power.   It is not really a true
energy form at all.   It is the controlling influence
which forms the plan for organising both matter and its
motion.

The effects previously considered, such as
telepathy, then appear as minor residuals of a major
driving force of the universe.   Without the psychic
element physical energy could not be structured to
provide systems of matter capable of building universes
or capable of motion.   **Psychic energy is:-**

The Intelligence Behind the Universe!

It is the number-crunching power of the Grid acting
as an ultra-fast computer.   Consequently psychic energy
is not a true energy form at all because it has an
abstract number-like quality.   Because of this it cannot
be measured in the manner of physical energies.

A point has now been reached at which an attempt can
be made to explain the whole spectrum of psychic effects
in terms of extended physics.   Before attempting to do
so, however, the solution built up will be summarised in
the next chapter.   A small amount of new material will,
however, be incorporated.

8   THE SOLUTION SUMMARISED

PLUS
DEFINITIONS OF TERMS

8.1   SUMMARISING THE SOLUTION FOR WAVE-PARTICLE DUALITY

Before looking into the way reported data can be explained it is desirable to summarise the solution built up by the argument developed in preceding chapters. Some new material is included, so the stoics who have not jumped a few chapters will still find it worthwhile to stay with this revision part. First let us be sure we have everything well-defined so that there is no room for ambiguity.

Light is a form of energy which could be defined as a physical form, entirely different from the psychic kind. Only physical kinds are as yet considered in physics. We will usually call it just "energy", understanding that the psychic kind is not involved unless specifically stated. In the new solution sub-atomic particles are at all times solid little objects built from energy. Nobody knows what energy really is since this is beyond our depth of understanding. However, the basic laws of physics which control the way energy behaves are well understood. So we use the term as if we know just exactly what it is. Physical energy, for example, is of the kind involved in the lifting of a weight. Indeed the weight of an object is the force exerted on it by the gravitational field of the Earth. This force, multiplied by the height through which it is lifted, is a direct measure of the energy expended in lifting. Energy can exist in many alternative forms

135

and, in principle at least, all forms are mutually convertible. For example, in a car engine the chemical energy locked up in the fuel is first converted to the energy of high temperature gases. This is actually a form of "kinetic energy", that of motion. This is because by burning the fuel the speed of molecules, of which the gases consist, is greatly increased. The engine then converts some of this "heat energy" to the mechanical kind. Then if the car it is driving is moving uphill, some of this will be converted to the kind involved in the lifting of a weight.

Einstein's famous equation, $E = m.c^2$, shows that energy "E" is equivalent to mass "m" multiplied by the square of the speed of light "c". Hence an object at rest and accorded a "rest mass" "m_o" can be considered as constructed from an equivalent "rest energy" "E_o". So objects at rest are made from rest energy and when they move an energy of motion or kinetic energy is imparted. These energy forms add up to a "total energy". The "m" in Einstein's equation, usually called the "relative mass" due to the history of its origin, is then equivalent to this total energy. This total really needs to include all kinds except gravitational potential energy (GPE) and the psychic sort. The exclusion of GPE is a new departure, the reason for which will become clear later. Hence all matter and its motion are somehow constructed from energy.

This description applies to electrons and the nuclei of atoms but the energy carriers of light, have no rest energy. They cannot exist at rest, being constructed of kinetic energy alone. As a consequence they can only travel at the ultimate propagation speed of physical energy- the speed of light.

Next, quantum theory shows all energy forms are discontinuous. Energy exists in discrete packages called "quanta". A quantum of light is a particle called the "photon". Countless billions of them propagate in streams when an object is illuminated so that we can see it, but they move in a complex way. They each carry with them oscillating force-fields of electricity and magnetism, even though they do not themselves carry any electric charge. These are "transverse" waves, which means that a sideways oscillatory motion is induced like that of a skipping-rope. The direction of this motion, as shown by the electric component, gives the direction of "polarisation" of the wave. A beam of ordinary light has photons snaking about in all sideways directions. A piece of polarised glass only allows those moving in a certain sideways direction to pass. It acts as a filter to provide "polarised light". Light has the same nature as radio waves. The only difference is that the frequency of waves is much greater in the case of light. All such waves depend on photons as carriers.

They interact strongly with sub-atomic particles called "electrons", which do carry charges of electricity. Electrons can be caused to accelerate rapidly by allowing them to fall in electric fields so that they can move in beams, spinning on their axes as they go. They form the active parts which create the pictures on television screens. Photons can be both produced and absorbed by electrons. Indeed it is the violent agitation of electrically charged particles which causes photons to be emitted. This is usually the way light and radiant heat are created. But the natural home of the electron is the atom.

Matter when subdivided to its limit is found to be made from atoms. Atoms are not the solid little balls they were at first assumed to be. Instead they are almost empty space. They have a centrally located and relatively massive nucleus carrying a positive electric charge and the size of the atom is governed by the negatively charged electrons which move around them in a complex way. The opposite electric charges produce a powerful attractive force which holds the electron close to the nucleus. A few atoms forming part of a crystal are illustrated in FIG.2.*

Until quite recently I considered scientific knowledge of this kind to rule out the possibility of a spiritual side to existence. I think the way my about-

*See Page 40

turn took place must be unusual. It resulted from a deeper study of physics!

This conversion arose from trying to work out a plausible solution to the mystifying phenomenon called "wave-particle duality" (WPD). It implied that an all-pervading network must exist throughout the universe, connecting everything to everything else as has been explained in earlier chapters.

WPD can be understood best by considering the famous two-slit experiment of Thomas Young illustrated in FIG.1. He passed a narrow beam of light through the two slits and showed that an "interference pattern" of stripes appeared on a screen placed further away. Wave patterns can be produced by throwing two pebbles together into a still pond. An interference pattern arises in the region where ripples spreading out from the two centres cross over each other. Consequently Young's experiment demonstrated light to have a wave nature.

The waves spread out from the two slits and where the two sets of peaks coincide they mutually reinforce, but where peak meets trough they cancel. In FIG.1 the places where "constructive interference" occurs are shown by short thick lines. These are where peak meets peak and trough meets trough.

But two discoveries were subsequently made which threw the entire scientific world into total confusion. The same pattern was built up when only one particle travelled through the apparatus at any one time! Worse still all sub-atomic components, known to be discrete particles, such as the electrons which buzz around the atom's nucleus, also behaved in exactly the same way as if they were waves.

How could a single particle go through both slits at once and interfere with itself like a pair of mutually interfering wave trains? This was the vexed question which caused so much head-scratching.

Many solutions were proposed, some of which assumed that particles were made from "wave functions" whilst in transit and only collapsed into particles on

observation.   (The term "wave function" is scientific
jargon which means the "heights" of all waves added
together and then squared.)   But none were consistent
with all the facts and paradoxes abounded as described
in Chapter 3.

The main purpose of this book was to develop a more
rational interpretation free from paradox i.e. any kind
of contradiction in either the initial assumptions or
predictions.   In the new solution which developed,
particles and waves are considered as totally separate
entities!

Particles are now always tiny objects made from
physical energy.   But in the new interpretation what
appears to us as a permanent entity must in reality be
a sequence of very short-lived particles.   They are
unstable and keep collapsing to leave the energy from
which they were constructed in a disorganised state.
Disorganised energy needs to be used as raw material for
the reconstruction of a replica particle having the same
momentum, velocity, spin and any other relevant property
such as polarisation, as its predecessor.   Such
requirements need a control system.   The point of
reappearance also needs some form of control system if
light or matter is to act according to the manner
observed.

Waves form the required control system.   They are
not made from either energy or anything else.   They are
of an abstract nature like numbers.   Wave-numbers need
to be added and squared at closely-spaced points and
allocated coordinates arranged on a three-dimensional
grid.   Hence they need to be manipulated by some form of
computing system.   The way waves can be represented by
numbers is described in detail in Chapter 4 and is worth
reading if the statement just made does not seem to make
sense.   A three-dimensional structure of criss-crossing
filaments needs to be postulated as existing in space.
This cannot be detected, because it is beyond the reach
of our senses.   It is fine-grained even on an atomic
scale.

The   crossing-points   of   filaments   form   natural

switches.   Now vast numbers of switches, acting in
concert, form the basis of all computers. Such a system
must exist and could have arisen spontaneously.   It
could have developed by some form of forced evolutionary
process as described at length in Chapter 6.   It also
needs to contain the means for particle reconstruction,
using as raw material the residual energy of previously
expired particles, whose disorganised energy has been
absorbed by the filaments.   At least 1,000 particle
factories need to be housed within the orbital of each
tiny atom.   Let us call this complete structure the
"Grid".   Although mind-boggling in its immensity,
considering that such a structure must extend for
billions of light-years throughout space, the concept is
not fundamentally impossible to imagine. It just means
that the deeper we look into the meaning behind
existence, the more wonderful and astonishing it appears
to be.

So the particles themselves are not waves at all.
They are simply controlled by waves which are totally
abstract in nature. Waves are not made from energy.
They are just numbers on the Grid. Where these wave-
numbers add up to their highest values with the total
finally squared, the new particle of a sequence is most
likely to re-appear.   Indeed a pattern can be built up
represented by these abstract quantities which
represents a probability distribution for organising the
places at which particles are to re-appear.

By referring to FIG.1[*] it works like this.   A
particle might go through one of the slits and some time
later find itself sitting on one of the short thick
lines shown just beyond the slits. It vanishes. The
Grid effectively throws a dice (random number
generator). An oversimplification can be considered to
make things easy. Each column of thick lines has
previously been allocated a reference number between one
and five. The next appearance will be one wave step
further from the slits but on any one of the five places
shown by the next set of thick lines. So we think of a
five-sided dice and choose a number at random. And so

See Page 36

the place of re-appearance is determined. The particle
lives for a short time and collapses. Then the
procedure is repeated. This continues until the
particle hits the screen at one of the five allowed
places. As more particles follow, jumping about
sideways from one allowable path to any of the others in
the same way, a pattern of speckles will build up.
After many particles have passed, the characteristic
interference pattern of stripes will appear on the
screen in accord with observation.

In this way particle-sequences really can travel all
allowable paths simultaneously. The model provides a
perfect explanation for the way a particle seems to go
through both slits at the same time. And there are no
paradoxes or other contradictions to worry about.

The foregoing description will apply equally to the
motion of any kind of sub-atomic particle. It could be
an electron or a complete atomic nucleus, for example,
or it could be a photon of light. The latter case is
the more complex because more than one wave is involved
at the same time. The photon will be controlled by the
abstract wave-numbers just described, but simultaneously
the electromagnetic wave of real physical energy will be
propagated. The two very different kinds of wave must
not be confused. The universe becomes ever more
mysterious.

Atoms need to be organised by similar abstract wave
patterns, since they too must be built from particles
which keep collapsing. The waves are now wrapped around
the nucleus of every atom. This occurs because each
electron possesses an electric charge which causes it to
be attracted toward the nucleus. The same force is
represented as an abstract quantity by the Grid, so that
the wave-pattern is affected and curls up on itself to
specify a ball shape. That this happens was predicted
by the famous wave-model of the atom derived by
Schrödinger. His equations described a probability
distribution, showing the places in three dimensions at
which the electrons were most likely to be found. They
were confined to ball- or lobe-shaped "orbitals". This

model has for many years been accepted as fitting the observed properties of atoms in a very exact way.

If lifetimes are governed by Heisenberg's uncertainty principle, then the electrons make about 250,000 re-appearances at randomly chosen places in the time they would have taken to orbit the nucleus as permanent entities in the manner of the Bohr description of the atom. FIG.2* can be considered as a picture taken of atoms using an exposure time very brief in comparison with Bohr's orbital period. Each electron has made several re-appearances at random positions. If the exposure time is increased more dots will be seen and when equal to the orbital period all would have merged into a diffuse cloud. In this way the atom can appear as a distributed fuzzy ball of electric charge. The complete atom is electrically neutral owing to an equal and opposite charge existing in the nucleus. If the time taken for disorganised energy to become absorbed by the Grid and jump across the orbital is assumed to be not less than 0.1 of the particle lifetime, then it is readily shown that the speed of propagation along the Grid filaments must be at least 10,000 times the speed of light!

This model for the atom fits in with that specified by Schrödinger's equations to perfection. Nobody has been fully able to reconcile his model with the concept of a permanent electron, which would be committed to orbits like planets going round the Sun. The hydrogen atom is the simplest, having only one electron. If the electron were truly permanent, then this atom would appear the shape of a thin disc instead of as a ball.

In addition "non-locality" or action at a distance is readily explained. A pair of particles, arising simultaneously like identical twins, seem to affect one another instantly whatever their distance of separation. The new explanation is that properties like spin or polarisation are organised to remain the same for all reconstructions. They do not exist in all possible states at once, in a state of limbo waiting to be resolved, as the Copenhagen interpretation demands.

See Page 40

With the new model non-locality does not really pose any difficulty.

The new interpretation therefore not only satisfies the observed phenomenon of WPD, it also resolves the difficulty of explaining the shapes of atoms and offers a simple resolution for non-locality. Previous interpretations do not provide such a comprehensive solution. Since also no paradoxes or unreasonable assumptions or predictions are involved, this new interpretation needs to be accepted as a serious contender.

According to the new model, however, the atom ought to radiate photons at a prodigious rate and this does not happen. There is also another question regarding the need for real electrical forces. Solutions for these difficulties were advanced in Chapter 7.

Some particles spontaneously decay into other kinds of particle after a brief life. For example a charged "pion", commonly seen among the cosmic rays penetrating our atmosphere, will spontaneously split in two. It is said to decay to a charged "muon" and an uncharged "muon-neutrino". In the new interpretation it is the **sequence** of pions which is broken. Its particle-sequence of real energy breaks up into a pair of other types of particle-sequence. The pair will carry away the same total amount of real energy and momentum as did the original sequence. One kind of particle-sequence is transmuted into other kinds of particle-sequence and all are wave-controlled. Hence what are already known to be unstable particles have another kind of instability superimposed on the primary sort introduced by the new concept.

Some further important implications of the new idea now require discusion. The Grid must offer a means of information transport as well as energy propagation at speeds of at least 10,000 times that of light. This is the "speed of thought" for telepathy which must be transferred directly along the Grid filaments.

The number crunching power of the Grid must be

inconceivably huge. This can lead us to a proper
definition for the term "psychic energy". It is not a
real energy form at all but an abstract computing power.
It forms the plan by which physical energies are
organised into the structure of matter.

8.2   CONSCIOUSNESS AS A SEPARATE ENTITY

Some effects, such as poltergeist activity, cannot
be explained without accepting the existence of an
invisible companion planet whose sub-atomic particles do
not couple with our own. This companion also needs to
support intelligent forms of life if a plausible
explanation is to be offered. Bizarre though this
requirement may appear, it is however completely
allowable by the extended physics of the new theory.
The selective coupling of one sub-atomic particle to
another via other particles of the vacuum called
"mediators" is required to explain, for example, the
electromagnetic force. To explain interpenetrating
systems of matter the idea of selective coupling merely
needs extension. Particles need to be able to recognise
a certain frequency band, interact with particles
carrying that label and ignore all others. In the new
theory these frequencies are associated with the wave-
numbers and are abstract quantities. This extended
physics allows several independent systems of matter to
be supported on a single Grid. Hence the idea that the
consciousness can exist as a separate entity from the
body is also supportable by physics. This is called the
"soul" in religious circles and so physics now comes out
as a supporting prop!. It is also clear that such a
system of interpenetrating universes could not have
arisen by happy accident. The fine tuning of the
physical laws involved could only have arisen as a
deliberate creative act. Some part of the otherwise
amoral Grid must therefore have evolved further to
develop a consciousness and a driving will. An
evolutionary process could have been involved as
considered at some length in Chapter 6.

The new theory therefore differs in a fundamental

way from established physics. As a result of history
the antipathy built up between science and religion
seems to have blocked the way toward this kind of
solution. Established theories all assume the universe
arose by accident and disallow the possibility of any
spiritual existence. If this approach is fundamentally
wrong, and it begins to look as though this is the case,
then established science could be stuck on a false
trail. A strange unholy alliance seems to have evolved
between science and religion for the support of the
purely materialist case. For as well as seeming to
support the contrary scientific view, strangely, some
religious sects also use it as a prop. It seems that
the established Church wants to go on as it always has.
It wants its congregations to follow by blind faith and
faith alone. It does not seek justification from
scientific proof or any other kind of proof for the
existence of any hereafter. Indeed it strongly opposes
any such proof. It seems afraid of proof.

To me this looks a totally absurd response. Without
proof, in the present climate of public knowledge, the
established Church can only wither and die. It **needs**
proof of the basic elements of its belief system and
these are common to almost all religious faiths. They
are simply the idea of a creative force behind the
universe and a spiritual existence continuing where
Earth-life leaves off.

Yet these things are compatible with extended
physics as we have already shown. There need be no
dividing gulf between science and religion. The two
disciplines need to appear as different aspects of the
same thing in total harmony and in collaboration with
each other. Then and only then can both revive. The
Church of England clearly needs a new look. Less than
2 1/2% of the population are said to regularly attend
its services. There has been much criticism of physics
in recent years. It is simply not producing value
commensurate with investment. So this discipline also
needs a new direction for its revival.

The existing solutions for wave-particle duality

seem to lean over backwards to avoid supporting any
possible creationist view.   They involve infinite
numbers of interpenetrating systems of matter at any
place, contravene the law of causality, or postulate
that universes are created by the combined minds of
human beings. Yet they fail to suggest how such minds
can be connected.   These minds, nevertheless, even
structure matter in the past as well as the present, in
order that they can exist!   Furthermore such older
solutions, despite extreme complexity, fail to provide
answers to all the phenomena presented by Nature. They
do not resolve the way in which Schrödinger's wave
equation can model an atom whose electrons are made from
permanent energy. They do not throw light on the way
non-locality works. The established explanation is that
space is so highly curved in higher dimensions that it
folds back on itself.   Then all points in space are
connected together in some kind of short-circuit. This
explanation is quite impossible to imagine because, as
far as we can prove, space can only have three
dimensions of length. And of course older theories do
not explain psychic phenomena.   The complexity,
implausibility and inadequacies of these interpretations
need to be remembered.

Then the new concept is unlikely to be discarded on
grounds of the sheer scale of numbers or other mind-
boggling features which are involved and cannot be
avoided. These simply mean that the universe is an even
more wonderful creation than we have ever imagined it to
be.

8.3 REAL AND VIRTUAL PARTICLES

In established physics the sub-atomic particles of
matter have been considered "real" in the sense of
having a permanent existence.   But others are also
necessary whose existence is temporary. According to
quantum theory "empty space" is not empty at all but
consists of a seething mass of "virtual" particles.
This is the case even in the space of high vacuum
existing between the stars and galaxies. Such particles

are invisible and each exists for only a fleeting
instant because it has no permanent energy. Their
effects are observable because some are the so-called
"mediators" which push sub-atomic particles about and so
produce the forces of nature.

Virtual particles, arising in unimaginably huge
numbers at random throughout space, live for a fleeting
instant and vanish. Some will impinge upon sub-atomic
particles of much lower density and are transformed by
a kind of catalytic process into mediators. These are
specialised virtual particles carrying tags so that they
can be selectively recognised by other sub-atomic
particles. When recognised they interact with greater
probabiltiy than when unrecognised and transfer their
momentum. The interacting sub-atomic particle then
experiences an unbalanced force in line with the source
of the mediators.

Some mediators are made from positive energy and
carry positive momentum so that a force of repulsion is
produced. Others are made from negative energy and
carry negative momentum. These account for the forces
of attraction. For example, the electric forces of
repulsion and attraction are explained this way in the
new theory. More details of the coupling scheme are
illustrated in FIG.6 and described in Chapter 10.

For the new scheme to work, half the virtual
particles of space have to be generated in positive
energy states with the other half negative. Despite the
unimaginably huge densities of either half, space
overall, therefore, acts as a fluid of zero net energy.
The particle generation needed to maintain this fluid
derives from constructional facilities distributed all
over the Grid filaments, which are also made as a
balance of positive and negative energy forms. The
fluid space, rather similar to air, can flow past the
Grid filaments as if it had an independent existence.
The matter we see is built from a tiny fraction of the
total number of particles comprising space.

The new theory deduced that real particles are not
real at all but have short lives, just like the virtual

particles of space. The difference is that the energy
from which they are constructed is real. When a
particle expires its energy is absorbed by the Grid.
Energy is then to be thought of as a mercurial fluid
able to instantly transfer to any other place in this
form. It is used for new production so that a real
particle actually consists of a series of
reconstructions based on the same piece of energy. They
are controlled by numbers on the Grid. They are not our
numbers, instead the numbers used are "wave amplitudes",
which can be added like our numbers and squared to
produce "wave functions" so that interference patterns
can be mapped. The Grid, therefore, needs also to have
a computer-like capacity, with the ability to add and
store wave-functions for the control of particles.

Fantastically complex though this structure has to
be, it cannot be dismissed as absurd or impossible. The
quantum description of space adds supporting evidence.
Physics has already accepted that some virtual particles
are complex structures similar to electrons. Many of
these act as particle factories themselves, throwing out
tagged mediators in all directions during their brief
lives as they fly. These tags enable them to be
recognised in a selective manner by other particles they
encounter. The latter then interact either strongly or
weakly according to a pre-arranged plan. It seems less
extraordinary to think particles arise from some hidden
constructional facility than that they just happen
without a cause as in established theory.

It is also satisfying to see that the new concept
provides a new kind of unification. The so-called
"real" particles which are observable have essentially
the same description as the virtual ones of space. In
older theories they are similarly constructed but real
particles are relatively permanent, whilst only virtual
ones have fleeting lives.

Finally, in order to provide a mechanism able to
explain attractive forces half the mediators of space
needed to be carriers of negative momentum. (Momentum is
defined as mass times velocity). This meant they had to

consist of negative forms of energy. A condensed object like a star had its concentrated net positive energy balanced by a tenuous halo of negative energy spread out over a huge volume of space. This halo consisted of the primary mediators responsible for the universally attractive force of gravitation. The two kinds of energy, positive and negative, are complementary. They are the Yin and Yang of both matter and space, absolutely dependent on each other for their existence. This is because at the beginning only nothingness existed. Only exactly equal and opposite forms of energy could arise from nothing to accord with a modified First Law of Thermodynamics. Then the same zero value of net energy of the universe exists at all times, including the transition across the creative event. The Grid structure is also therefore limited to a construction from filaments of balanced positive and negative energies. The crossing-points form natural switches and these form the basis of the computing system.

Matter and space though continuously regenerated from the grid filaments are so contrived as to enjoy a relatively independent existence. They flow past the filaments in the manner of a fluid. As shown in Chapter T.S.2, this fluid cannot offer resistance and so cannot be in any way viscous. It must be a superfluid like helium. Also as described in Chapter 5 and further studied in Chapters T.S.2 and T.S.3, this fluid needs to be slightly compressible. It is a compressible superfluid. Yet it has very special properties. Space exists as exactly balanced halves of positive and negative energy states and so its net energy density is zero. It has negligible inertia. A tiny fraction of its particles constitute matter owing to the selective coupling properties between certain types of sub-atomic particle. Owing to this coupling, apparently solid objects are simulated, whilst in reality they flow as a fluid past the Grid filaments.

Gravitation would tend to affect the Grid filaments just like matter. It is therefore necessary that virtual particles be fired from them in a counter direction to impinge on those of space, so transferring the force to the fluid which established physics calls "space".   In this manner matter can clump by gravitational force without affecting the Grid.

This model fits the data we have available. Since no other existing model does, extraordinary though it may appear, it needs to be considered as a serious proposition.

In order to make sure everything has been grasped properly it will be worthwhile summing up the points of major importance. First let us collect the definitions of terms used.

8.4 SOME DEFINITIONS IN PHYSICS

ENERGY   (Or "physical" energy measured in joules). Energy is the building substance of both the matter and motion of all objects in the universe. The energy of matter, the "rest energy", which is the amount measured when an object is not moving, can be added to any energy of motion, the "kinetic energy". Hence when an object is in motion its "total energy" is the sum of the two and so is higher than the rest energy.

There are + & - kinds of energy, irrespective of whether they carry electric charge or not. But if a + form hits a + form it knocks it away, yielding a repulsive response, and if a - form hits a + form it pulls it back, producing an attractive response.
(Only + energies are allowed in established quantum theory)

THE QUANTUM VACUUM

When air is pumped out of an enclosed space the number of molecules flying about inside gradually reduces. Ultimately it is theoretically possible to produce an absolute vacuum, in which case no free molecules exist inside. However, according to quantum theory this does not mean that nothing is present. In quantum theory the vacuum is filled with a seething mass of "virtual" particles, which are, however, totally invisible.

VIRTUAL PARTICLES

Are sub-atomic particles of the quantum vacuum. They arise spontaneously from nothing without any permanent energy. The energy needed for their construction needs to be "borrowed", like a debt needing to be repaid an instant later. Hence they arise from nothingness and vanish again to nothingness. The name "virtual" means lack of reality energy-wise.

MEDIATORS

Are the specialised virtual particles of the quantum vacuum which produce forces by hitting objects we can sense. Mediators cannot be sensed directly. They are generated from the surfaces of sub-atomic particles with special tags so that they can be recognised by others they may hit.   Whether they interact or not is determined by these tags. Specialised mediators allow matter to be structured, so that particles of the nucleus feel the strong force, for example, but the electron clouds around them do not.

This is all pure quantum theory except that negative energy states are not yet accepted to exist in establishment physics. This is one of the new features.

PSYCHIC ENERGY (Again this is a new definition)

This is where there may still be some confusion and the cause is that the word "energy" is being used in a different sense from that used in physics. We really need a new word for this. My definition of psychic energy is the computing power of the Grid which was shown to be required, permeating all space. It is not energy of the kind from which anything is made, rather it is an abstract form like a set of numbers. A combination of numbers can be used for control, so this is more a power of information as in a control system.

This means that when people talk about "force" or "energy" fields in regard to dowsing or psychometry, the physical energy of matter and motion is not relevant and must not be equated with the psychic form, which is the one operative in such cases.

THE UNIVERSE AS PSYCHIC AND PHYSICAL
ENERGIES COMBINED

Yet in another sense the two can be combined but this is not a straight arithmetical addition, because they do not have the same units. Only things which have the same units can be added together. For example it is not possible to add 10 seconds to 2 feet, because one has the units of time and the other the units of distance. Indeed the units of physical energy, whether positive or negative, can be measured in the units of joules. But psychic energy, being an abstract quantity, has no units. The latter is simply a control system used in this new theory to locate the places where real energy, that of particles, is to be allowed to appear. In the same way the energy or power of mind is the psychic form. Again this is information and is abstract.

To structure matter according to this theory, both forms appear. The psychic form arises from "wave functions" stored in the memory banks of the all-pervading Grid as numbers to provide an information system and this specifies where the physical energy is to be located. If preferred, this could be called "real" energy.

Hence in the new theory psychic energy is not reserved for other planes of existence with real energy being allocated to our level. Instead all levels of existence, assuming that a set of interpenetrating universes actually exist, are built from a subtle combination of both psychic and real kinds. Psychic energy provides the plan by which the real energy is structured, so that matter appears and acts in the way it does. The real energy then constitutes the building-blocks. But real energy alone would be disorganised and so could not exist as the matter we observe.

ELECTRIC CHARGE

Charge is connected with the electrical forces of attraction and repulsion and in quantum theory these forces are produced by "mediators".

There are two kinds of electric charge, + & -

+ & + produces mutual repulsion
- & -   "   "   "
+ & -   "   "   attraction.
To produce these responses the particle has to throw off
both + & - forms of energy as mediators and one or the
other has to be recognised for interaction to arise.
The "spin-coupling" scheme shown in FIG.6 provides a
possible explanation and is discussed later. (Again this
is not established quantum theory)

So there are + & - forms of physical energy and + & -
forms of electric charge. The two cases must not be
confused, though they are related in that for forces to
arise between charged particles a transaction of
particles made from one or other kind of virtual
physical energy is required (Again a new feature).

MORE ABOUT ATOMS

Atoms are constructed from minute sub-atomic
particles. They have dense and relatively massive
nuclei, surrounded by very light electrons. The
hydrogen atom is simplest, as it has as its nucleus only
a single proton and a single electron moves about it
confined to a ball-shaped orbital. The electron sparks
in and out of existence in a random manner anywhere
inside this ball of space controlled by psychic energy
as numbers patterned on the Grid and in so doing defines
the size of the atom.

So a rather complex system obtains. The
electromagnetic force, a combination of the electric and
magnetic kinds, has to be present in order to bind atoms
together but only the abstract model of this force is
needed to confine electrons within a tiny ball of space.

Atoms can stick together because of this
electromagnetic force. Magnetism arises when electric
charges are in motion. The electrons are spinning and
also travel during their brief appearances and so
magnetic effects are present. But although the atoms
stick together they do not interpenetrate very far
because the mutually repulsive forces of the electrons

prevent this. Consequently they appear hard and solid to us, whereas in reality they are almost empty space and their hardness is an illusion.

Atoms are therefore not solid at all. They are mostly space. All complete atoms are electrically neutral because the positive charge carried by the nucleus is always balanced exactly by the negative charge of the electron cloud.

It is quite conceivable that other forms of matter could exist made from the same kinds of physical energy and organised in the same way by psychic energy as information on a common Grid. They would be based on the same fundamental laws of nature as our own, but if their mediators were so arranged that they did not couple with our sub-atomic particles, then they would not be detectable. There is so much space inside the atom that all such other forms could mutually interpenetrate.

Our horizons should now have been extended by new insights. Let us therefore explore the universe from a new angle. In the next two chapters we will investigate what many people say is inexplicable.

9 EXPLAINING THE UNEXPLAINED

or

EXPLORING THE PARANORMAL

For hundreds of years reports have persistently appeared of "supernatural" phenomena. Colin Wilson(125) gives an interesting analysis of some of these in *Poltergeist!*. In addition there has always been speculation about the validity of water-divining, black magic, white magic, spiritual healing, the sighting of ghosts, "out of body" experiences and telepathy. And this is by no means a complete list.

In previous centuries such things were discounted as hallucinations or deliberate fraud because they could not possibly be fitted into any scientific theory. But then the workings of the universe were being reconstructed before the implications of wave-particle duality were appreciated.

But is it scientific to discount reports simply because they cannot be explained? Better to file them for future reference in case they have some substance. Clearly, with a Grid connecting all things plausible explanations for such phenomena can be expected. Perhaps the time has come to use these reports as input data!

A personal experience always seems to me the most convincing and so it seems appropriate to make a start by recounting one of these.

9.1 PSYCHOMETRY

My wife Margaret made the crazy suggestion that we join a development circle for beginners to see if we had

any psychic powers. I could not offer a logical
objection, having already accidentally discovered that
I had a theory which predicted psychic effects. This
had arisen as spin-off from an extension to Newtonian
physics. Yet it seemed absurd to think this spin-off
into the paranormal could actually have a true bearing
on reality. But I had to go along with the idea.

We both thought the first session rather childish
and were not impressed. We seemed to be going back to
childhood imagination games. We were pretending to open
our imaginary heart chakras and throat chakras finishing
up with our crown chakras. Then we were bringing down
imaginary beams of light and surrounding ourselves with
them. We were walking down avenues of trees and
climbing down steps, then going through imaginary doors
to sit in private imaginary refuges. All very silly.

Still we went along next week. This time, after the
preliminaries, we had to place our hands in turn on the
shoulders on the next person in the circle to see what
we could pick up. I knew this would'nt work and nothing
happened, at least for me and for the only other male
participant. But all the ladies got something! Each
described seeing pictures which all turned out to have
relevance for the person being touched. Despite their
emotional and supposedly irrational nature the female
brain seemed to work where ours did not. Perish the
very thought! However, we enjoyed it much more this
time. It did not seem so foolish when the pretending
could produce results.

I asked Vic, who helps his wife Barbara in running
the circle, a question to which I very much wanted an
answer. I had heard that sensitives could obtain
impressions of people from objects which belonged to
them. Was this true? I asked. I was interested
because my theory offered an explanation and I was
seeking experimental confirmation.

To explain the puzzling scientific phenomenon of
wave-particle duality, to which I have referred in
earlier chapters, a computer-like "Grid" had to exist.
It was composed of filaments fine-grained even on an

atomic scale yet, criss-crossing one another, they
pervaded all space. They connected everything to
everything else yet were beyond our visual perceptions.
The number crunching power of the Grid provided a
control means for real physical energies both positive
and negative. It provided the plan by which physical
energies were organised to structure the matter of our
universe. This computing power defined the true nature
of the term "psychic energy". Like man-made computers,
the Grid needed to incorporate a memory bank but this
would be unimaginably huge by our standards. If
sensitives could address this memory, then they were
seeing the tiniest tip of a huge iceberg of information.
It would not be surprising if the memories of all
experiences were stored on the Grid inside atoms. If
therefore there was any truth in the rumour I had heard,
the human mind was able to penetrate right into the
hearts of atoms. If the memory of the Grid could be
read then valuable confirmation of the theory would
exist.

Vic's answer was to take my watch. What he told me
about my past was highly relevant. Some of the things
he said were not totally complementary and I will not
repeat them! When he ended, Barbara and Vic decided
that next week's sitting would be devoted to
psychometry. Each of us had to bring along something
personal.

We both thought that Barbara and Vic would read
these for us. But not a bit of it. We were led through
the preliminaries as before by imagination. Then **we** had
to do the readings ourselves working in pairs! I just
knew nothing was going to come out of this. I was to
read a medallion Vic passed to **me** if you please. I
decided I would just have to be honest and tell Vic that
what he was asking was totally impossible, I would tell
him all I could draw was a blank. I held his medallion
for a while and tried not to think about anything as
instructed.

Nothing happened for about half a minute. But then
I suddenly became aware that my hand was hurting and

felt stiff.   I transferred the medallion to the left
hand. This hand hurt and went stiff and the right hand
relaxed again.   So I transferred it back again and the
stiffness and pain moved with the object.   I was
astounded.   Vic said "Now you know what rheumatoid
arthritis feels like. I have it in both hands." He had
not mentioned this before.

All the ladies got something as well. But they all
saw pictures.   And again they all turned out to be
relevant.

This really is the most amazing thing. For me it is
most exciting because it offers valuable confirmation of
my scientifically based theory.   Yet although it
predicted and explains psychometry, my mind still found
it hard to accept as true.

Long held prejudices die very hard!

From this experience it seems to me that the
parapsychologists using accepted scientific methodology
should take note.   They are looking for so-called
"rational" explanations of the paranormal and should ask
themselves whether their methodology is appropriate.

But now let us review data from other sources.

9.2   TELEPATHY

With a Grid coupling all things there is a channel
for information which ought to enable people to see
without the use of eyes or hear without the use of ears.
It cannot be said that because we have no such ability
the idea of a Grid existing is flawed, however.   It
could be that this sense is deliberately suppressed to
avoid overload.   If all the information from the
universe were not filtered out we would be swamped.
However, it is clear that the concept of telepathic
communication cannot be held to be impossible on
scientific grounds.   It could be explained as a
controlled leak in the filtering mechanism.

In a private communication Mary Harrison, author of
many books on the paranormal, yet very sceptical of its
reality, gave an account of a private interview with Uri

Geller. She said he asked her to think of a picture
which he would try to draw. She concentrated on a
shamrock with heart-shaped leaves. He produced a result
which was very nearly correct. He obtained the outline
shape but added a heart shape at one place. He told her
he looked at an imaginary television screen and watched
pictures arising upon it. Any which lasted more than
six seconds would, he knew from experience, be close to
that in the thinker's mind. This is surely a convincing
demonstration of telepathy.

9.3   HEALING

If a way could be found for programming the Grid,
which must be mostly amoral, then quite magical effects
for good or harm could be worked. Since the Grid has
all the information of everything stored within its
memory-banks, it ought to know how to restructure any
damaged cells of living creatures.

The idea of healing by programming the Grid is
therefore scientifically possible. The placing of hands
by a healer near to a sick person's body could not by
itself achieve anything. But by the will of the mind
the computing power of the Grid could be brought to bear
for the restructuring of damaged cells. This would be
carried out not only atom by atom, it arises at the sub-
atomic level. This explains the psychic power of
healing according to the present theory.

Heat is said to radiate from the hands of the
healer. There may be some physical heat but it is
possible that there is an associated form of a psychic
kind which would not register on a thermometer. This
explanation is probably incomplete because all the
healers I have met claim to be assisted by unseen
"guides". We will look for an explanation to this
extension in the next chapter.

That spiritual healing is a real effect can be
judged from its increasing use in hospitals and by the
established Church. If all results were uniformly
negative its increasing use would not occur.

The evil of black magic cannot be rejected as an

impossibility. It could work by the same mechanism. It seems that for every positive there is a negative side which needs to be remembered and the necessary precautions taken against misuse. Both good and bad effects could act at a distance by transmission of programming power through the Grid itself.

According to information received it is important that some article belonging to the healing target be utilised. This is held by the absent healer before the healing power is discharged. The advantage of this, according to the present theory, is the assistance given to the Grid to help locate the target. The information of the memories will be transferred to the Grid via the route of psychometry acting in reverse. Then the Grid can search its memory banks for a matching memory, then discharge its matter re-structuring power at that location.

9.4 DOWSING

Since the Grid has memories of everything, the secret of programming would make possible the prospecting for minerals or water. Some people claim to have this gift. Do they use a hazel twig as a programming device when dowsing? Uri Geller possesses the power to an extraordinary degree and has demonstrated it in locating crude oil deposits. I am told he has made himself very rich this way. If this is true, then divining cannot be dismissed as fluke. Some further evidence came to me during a telephone conversation. Mr Delgardo, an electro-mechanical engineer, told me he found he could manage better using his hands alone. No twig or other implement was required.

The curious information he imparted was that he had discovered the existence of energy fields at the sites of circles of flattened corn. He means psychic energy, of course. In a book he wrote as co-author with Andrews(109) it is stated that a compass needle began to rotate by itself when placed in the centre of a corn circle. This kind of observation needs to be confirmed because it is to such strange effects that science needs to look.   The alternative explanation is deliberate hoax.

9.5   MIND OVER MATTER

Uri Geller is also famous for his amazing spoon- and fork-bending powers.   These have been amply demonstrated and verified under stringent laboratory conditions. They show the power of mind over matter.   Again the Grid hypothesis offers a plausible explanation.   The Grid manipulates wave functions to control the regeneration of sub-atomic particles, so that the illusion of permanent atoms is created.   Uri Geller possesses enhanced powers for programming the Grid.   He is able to override it and so cause it to fail to maintain its natural matter-structuring capability.

He has of course met with fierce opposition from the Magicians Circle.   It is not in their interests to have it accepted that such effects are truly paranormal.   I watched James Randi on television attempting to copy the spoon-bending feat represented as a conjuring trick.   He did not succeed.   He had to cover the spoon for an instant and then allow a bent spoon to appear.   Uri Geller showed the spoon in full view all the time and the bending was a gradual process.   Yet he only stroked it lightly with one finger.   There was no way he could have caused the bending by force.   Some children have displayed the same ability.   These claims have also been thoroughly checked.

In a lecture given here by George Meek, an American who has researched such matters for many years, spoon bending was touched upon.   He said that spoon bending parties are regularly held in the U.S.A.   They first create a highly emotional state by creating the party spirit.   A highly emotional state is apparently the key to success.   When this state is judged to have been achieved, everybody attempts to bend one of the spoons brought along.   He said that about twenty-two out of twenty-four are bent.   This represents an astonishing rate of success.

The fact that it can happen seems less extraordinary when a plausible scientific explanation can be advanced, though why emotion is involved is not yet clear. Perhaps this has something to do with the ability of the mind to reach down into the Grid to effect programming.

9.6 THE APPORT

One of the most interesting possibilities allowed by the new extension to physics is an instantaneous means of transport known as the "apport". For example, an egg could reside in a refrigerator. It exists as energy organised by the wave functions patterned within the all-pervading Grid. Programming could allow the wave-function describing the egg to be copied exactly in digital form at some instant but transferred to another place specified by coordinates. There would be nothing visible at this time but then, the procedure accomplished, the wave function of the egg within the refrigerator would be deleted, leaving its net energy disorganised to oversaturate space at that point. Instantly the energy would transfer along the Grid filaments to the newly-organised position, regardless of any intervening matter. It needs to be remembered that atoms are almost entirely empty with the grid filaments passing through them. Disorganised energy would be able to flow along the filaments through what to us seems solid matter as if it were not there! Only when organised by psychic energy into the form of atoms will the constuctional material, energy, be obstructed by other atoms.

A new egg would appear in empty space, to be immediately acted upon by the normal forces of nature, including gravity. It would fall to the floor with a splash. A pointless use of a valuable power. In fact the example chosen is taken from Colin Wilson's book. It explains a recorded event!

If the secret of accomplishing such a feat could be discovered it would open the way to exploration of the Galaxy. Rather like "Starship Enterprise". But instead of using rocket power to reach the stars and then

beaming down occupants to planets the entire spaceship
would be beamed to within the vicinity of a star. It is
very unlikely that this will ever be possible, of
course, but this way offers the only hope of exploration
outside the solar system by interstellar travel, so vast
are the distances involved. An example can show just
how impossible rocket flight would be for interstellar
travel.

If a conventional rocket is considered, using an
advanced propellant capable of an efflux speed 10 times
that obtainable with known liquid fuels, and if a
payload of 10 tons is to be delivered to the nearest
star, and if a fuel supply equal to the entire mass of
the Earth is to be allowed, then a transit speed of only
0.0015 of the speed of light would be reached! It would
take 1,800 years to get there.

It is just possible that more advanced civilisations
living on planets circling stars somewhere in our galaxy
could have discovered the secret of the apport. We
might look out for possible extraterrestrial visitors
who have used this mode of transport. Reports of flying
saucers cannot therefore be discounted, though they do
admittedly seem fanciful notions. The reports of UFOs
which have been given, seem to accord with the above
description. They have also been reported to suddenly
vanish. Such sudden disappearances are also explicable.
They have apported to a new location!

They could be made from exactly the same system of
matter as our own. Then they would be visible and would
appear solid to us. They would produce shock waves at
supersonic speeds just like our own aircraft. In
addition landed craft would cause depressions in the
soil because of their weight.

Numerous mysterious accounts have appeared in the
press and journals of flattened discs of crops in
cornfields. Most have small satellite discs arranged
symmetrically around them. These, not surprisingly,
have been explained as hoaxes produced by a team of
strange vandals. This seemed the most reasonable
possibility until Mr. Delgardo, showed on television,

by dowsing, that these rings produced signals similar to
those diviners obtained when looking for subterranean
water.   This finding seems to rule out the vandal
hypothesis.   The UFO alternative needs, therefore, to
remain as a possible option until some other explanation
is proved.   Why UFOs leave "force-field" traces we now
identify with psychic energy still remains an unsolved
mystery, of course.

Several months after this section was written some
very elaborate designs appeared in Wiltshire cornfields.
They aroused interest worldwide and so we went to see
one of these near Alton Barnes.   The following is an
extract from a submission to the competition advertised
by the "Sunday Mirror" which offered £10,000 to the
winning explanation.   Abbreviated it read:-

SOLVE THE CORN RINGS MYSTERY.

I aim to show that a solution to this puzzling
phenomenon might be found by using a new extension to
physics.   First however it is necessary to assemble the
facts.

The first corn circles were simple in shape.   For
these the idea of random effects producing whirling air
columns might seem plausible.   They could flatten corn
in the direction of airflow.   Other advances have been
made on this idea by adding electric charges to the
swirling air.   Then this mass has been assumed to form
into balls which fall and spread out on impact with the
ground.   But it is clear that over a period of several
years the shapes have increased in complexity.   The
latest shown in the excellent picture, published in the
"Sunday Mirror" on 22 July, is the most complex yet.
There is absolutely no way such simple explanations can
any longer be regarded as plausible.

The idea that they are all elaborate hoaxes also
seems implausible.   Observers have reported seeing other
circles form, hearing humming noises.   At night some
have reported seeing glowing balls falling to earth.
Only in the morning have the circles been observed at
the place of impact.   Since dowsers have reported

finding psychic energies it seems prudent to keep options open.

However a new theory regarding the structure of matter depends on the existence of an organising intelligence behind the universe. Clearly, for the creation of elaborate shapes, some form of intelligent control is implicated which may not derive from people of this planet. The charged whirling air columns provide a physical explanation. These, however, might be tools which are deliberately manipulated. By this means the intricate key shapes with precise sharp corners could be produced.

We need to take a microscopic look even below the structure of the atom to solve the mystery. In the new theory matter is built from particles of physical energy but these are controlled by an intelligent substructure or "Grid". The computing power of the Grid defines what "psychic energy" really is. Without this information content, physical energies could not be organised into atomic structures.

Also predicted by extended physics is the possibility of systems of matter which are invisible since their sub-atomic particles do not couple with our own. They can interpenetrate our matter without being detectable by us, yet are controlled by the same underlying Grid. Discarnate intelligences are permissible by the theory, associated with interpenetrating systems of matter. If they became aware of our existence and wished to communicate, then they could do so only with great difficulty.

The evidence suggests such intelligences may exist! Not only that. They may have found a way to directly affect our atmosphere and for several years have been experimenting, gradually improving their techniques. Are they trying to tell us something?

My wife and I visited the circles at Alton Barnes last Thursday (19 July) and lay in the centre of a flattened corn circle to see if we could pick up any impressions. We did not. However we have been attending a "circle" whose purpose is to develop psychic abilities and we have not yet developed very far. The same evening we attended our circle to be chided by Vic, a sensitive, for not bringing back a sample. However he decided it might be worth trying to pick up something from us at second hand. He held my wife's hands and soon reported feeling impressions.

What follows will seem quite absurd and it should certainly not be accepted unless corroborated by others. It is worth reporting just in case it turns out to be of value.

Vic said we would not be able to relate the entities concerned to life forms because of their shapes. His answer was literally a lemon. They had no limbs, looking like something between a lemon and a melon. He also picked up a repeated message saying that we should look eight feet down below the surface. There we would find a changed structure which would be inexplicable by our present knowledge.

Now there is no reason why all life-forms should have limbs. If discarnate intelligences are able to programme the grid directly by thought, as we do our computers using fingers, then they could make anything they needed without having to cut or carve the way we do. Therefore no justification exists for discarding this information on the grounds of its apparent absurdity. Yet on the other hand we cannot be gullible and just accept it. It is necessary to maintain open minds. There is a crying need for a dig. So how about organising one?

At Alton Barnes a row of about nine circles lay in exactly a straight line and some were joined by accurately shaped lanes about eight feet wide. Some days later another even more elaborate design appeared near Silbury hill. It seemed to be pointing somewhere and so a supplement was sent to ask for information. Abbreviated this read:-

WHERE ARE THEY POINTING?
The directions of these new lanes, do not follow the

tractor tramlines like those at Alton Barnes. They seem
to be pointing somewhere. One of three possibilities
must apply and so the matter needs to be explored. They
are:-

1) They point in random directions so the cross-over
points of different formations will be at random.   In
this case it can be concluded that there is no
significance in the direction of these lanes.

2) They are parallel. This would be puzzling.

3) They all have a common cross-over point. This
would be highly significant.   It would prove an
intelligence was behind it all, such as that described
in my first submission. It would also indicate that
something unusual would be found at the point indicated
and this would need investigation.

With these two as-yet unexplored proposals we will
leave the corn circle puzzle. If they do all ultimately
prove hoaxes, carried out worldwide, then they must be
the most organised of any ever known, being carried out
at great trouble for no obvious reason whatever. The
UFO could be connected with the mystery, however, and so
it may be worthwhile taking a further look at these.

UFOs and their extraterrestrial occupants need not
exhibit any kind of paranormal behaviour. They might be
tougher than us so that greater accelerations in
conventional manoeuvres could be permitted but not to an
extraordinary degree.

Which means that a second kind of UFO might exist,
because some extraterrestrials appear to accelerate at
unbelievable rates. The ones which do would have to be
made from a different kind of substance. They would
have to consist of the same basic energies as us, since
this is the universal substance of all matter and
motion.   However, their atoms could be arranged
differently. Their atoms could be formed with exactly
equal and opposite energies. In this case their matter
would have no inertia and extreme accelerations would be
possible. This matter is studied in detail in the next
chapter.

Consequently speeds close to that of light could be reached within a few seconds. But travel at superluminary speeds would not be possible this way. It is just possible that their sub-atomic particles could couple with ours. Then they would seem solid to us because their matter-systems would not interpenetrate ours. This possibility does not seem very likely, however.

What is more probable is that exactly balanced systems of matter would not couple with ours. Nor would their photons couple with our electrons. Photon-electron coupling is necessary for vision, because an effect similar to that of photoelectricity is involved, taking place in the retina of the eye. It follows that such matter would exist in the same spatial dimensions as our own but could interpenetrate ours without resistance. And it would be invisible. It is possible, however, that visibility might be conferred under exceptional circumstances by a system of transient matter coupling. Some of our matter could coat theirs, to render it visible by light reflected from such an interfacing cover. It would demand a covering substance having exactly the same atomic spacing as the substrate.

Some of the evidence from UFOs is also supportive of this explanation. Observations have been reported of UFOs flying supersonic without producing shock waves and diving into the sea without making a splash. These observations are consistent only with such an interpenetrating system of matter, able to pass through ours without resistance.

This is not impossible but the matter will be dealt with in more detail in the next chapter.

It would probably be much easier for people made from balanced energies to travel by the apport method than for the unbalanced kinds like us. Their net energy being zero, there is no need for energy transport at all. Mediumship has reported that people in our companion universes travel "quick as thought" by thought alone. This suggests that by normal thought processes the Grid can be programmed. This conclusion is

supported by the explanations given for spoon bending
and spiritual healing.

9.7 A NEW KIND OF ASTRONOMY?

For humans to explore the Galaxy it would seem safer
and more feasible to use the Grid merely as a system of
information transfer. Since the Grid will allow speeds
of transmission many thousands or possibly millions of
times the speed of light, it should be possible, in
principle at least, to create a new form of astronomy.
Its feasibility depends on acquiring an ability to
connect up in some way with the Grid. Once this has
been achieved it will be possible first to measure the
speed of propagation of thought-waves through the Grid.
If measurements are carried out in different directions
it would be possible to determine the speed of the Earth
relative to the Grid.

9.8 CAN MIND-POWER PROGRAMME THE GRID?

It is becoming very clear that our minds have an
important influence. In this theory they do not
actually structure the universe we observe, as in the
Everett-Wheeler concept, but they can have an effect.
This effect can be overriding under certain conditions.
It cannot provide the primary cause of existence in a
theory based on common-sense however, because the law of
causality needs to be observed. Cause must precede
effect.

It does look as though our minds are linked to that
of the Creator; perhaps we are a part of the universal
mind. The Creator could have divided the original
single mind into countless sub-minds distributed
throughout the animals and humans of the entire
universe. They would all be interlinked by the Grid to
maintain a unity at a deeper level.

A bizarre thought emerges of combatants in mortal
conflict. At the upper level they are antagonists
trying to destroy each other, yet at a more basic level
their minds are cooperating with the totality of all
minds and continue to do so until the moment when one is

extinguished.   But even then only the Earth body is
lost; the mind survives.   So both can continue to act as
components of the Universal Mind!

It seems very likely that such an arrangement has
been deliberately organised so that all kinds of
experience, both good and bad, can be transmitted and
stored for some basic purpose, to provide the Creator
with input stimuli and to permit in return the
expression of creativity.

From these considerations it does not therefore seem
unreasonable to consider that prayer may be a real
force, able to be used for good or ill at the desire of
the individual.

9.9 PRECOGNITION.

The   explanation   of   precognition   presents
difficulties. Others have suggested that there is no
such thing as time in an absolute sense, so that people
in another universe can see our future as if laid out on
a map. Now time can be defined as a sequence of events,
one happening after another. Without time there could
be no motion. Furthermore if every action to be made in
the future has already been settled by the past-
present- future time map, then there can be no free
will.   Hence this fixed map idea is difficult to
justify.   Yet hard evidence for precognition exists.
For example, one little girl kept dreaming she would die
covered by a black cloud which also smothered her
friends. She kept telling her mother about this for
several weeks before the disaster occurred. Then she
died, as she knew she would, smothered in the mud of
Aberfan.

It seems there could be two kinds of precognition.
One is the warning kind. In this, if avoiding action is
taken, the predicted event does no harm. The future
cannot therefore be fixed because a different sequence
will arise depending on the freewill of individuals
acting in the present. The law of causality would
otherwise be violated. The previous example is clearly
of the warning type. If the little girl had played

truant, the mud from the coal-tip on the mountainside would still have destroyed her school but she would not have been there. The Grid explanation is that all data about everything is stored in its memory-banks. Its computing power will be able to predict what is likely to happen just as our computers can predict the weather from existing trends. It would know that instability was increasing and would be able, within limits, to determine the probable time of earth-slip. Telepathic messages could be conveyed during sleep, the time of greatest receptivity, when brain activity, which could interfere, is at its lowest ebb.

There could be another type of precognition based on a future which is certain. This does not mean time does not exist, however. In this case some "will" in control of the Grid has decided that a certain sequence of events is to occur. Then, wielding its immense power, the Grid makes sure that this sequence does happen. Again the Grid might transmit telepathic messages, so that clairvoyants could predict such a future event, even centuries into the future. It is difficult to see, however, what purpose this information could serve. None perhaps. It may be that such information transfer is accidental.

This is the common-sense solution. It maintains that time flows just as it seems to do. It would seem to operate at different rates in different systems of matter. But events must occur in sequence and this means time progresses onward forever. There is no mystery about it at all.

9.10   RE-CONSTRUCTS

One final piece of paranormal input data needs to be considered. Very occasionally reports of solid structures have appeared which belonged to a past age. Complete houses have been described which were not phantoms, because they felt absolutely solid, yet they suddenly reverted to their normal state. Some have suggested time-warps as the explanation. But these are disallowed by the present theory, which permits only of solutions based on the law of causality.

The Grid, however, provides an alternative explanation. The huge capacity for information storage offered by a computing system based on components which are minute, even compared with the size of atoms, is beyond comprehension. The Grid can therefore store memories as wave functions of every stage and every moment of history. Occasionally somebody in a companion universe, having powers of programming which dwarf those of even Uri Geller, might decide to re-create an object from times past. The energy requirement can be derived from existing matter and redirected according to the old plan of stored wave-function-numbers. Whilst it lasted it would seem just as real as the original. Some kind of cosmic law we know nothing about may operate to force such re-creations to revert to normal after a short lapse of time. The normal state, to use computer jargon, is the "default" condition.

This explanation does, however, draw attention to one problem. Such manifestations have always arisen exactly on the site of the original building. This implies that the Grid filaments do not move relative to solid matter but are trapped inside the structure. To sensitives matter which is transportable clearly carries with it memories of the past. They are able to describe a deceased person by holding a lock of that person's hair, for example. This is psychometry and I know from personal experience that it works. It is difficult to see how bits of matter containing their own Grid structure can be moved without breaking the Grid filaments.

It could be that they are broken continually but then re-form, so that the interconnection of everything to everything else remains intact. In this revision of the Grid construct the filaments would only remain intact when fluids pass through them. When solid objects move, the filaments break and move with the object. Clearly this could only happen in one universe, so any others could not carry memories along with their matter-systems.

And this does not seem a very satisfactory state of affairs because, as will be shown in the next chapter, it is very likely that interpenetrating matter-systems exist.   The bits of Grid moved in our universe would upset the programming in others.   An alternative explanation therefore needs to be sought. One could be that the Grid filaments remain intact but the memories stored as switching sequences are dragged along with an object which is moved. Only a relatively small number of the switches would be involved, leaving the rest spare or for the programming of other universes. This seems a much more satisfactory arrangement.

In this way, by study of so-called paranormal data inputs, much can be deduced of the secrets behind the universe we appreciate by our five senses. Since the Grid concept can provide rational explanations, such data ought not to be regarded any longer as paranormal. It can now be regarded as natural since physics is extendable to bring it within its orbit.

To the extended view of physics portrayed in this book all data input is normal.

Many sensitives claim, however, to sense the existence of intelligences which do not belong to our system of matter. It is therefore necessary to explore this area in a scientific way to see if the Grid concept can offer any rational explanation. The word "rational" now, of course, no longer means limited to the single system of matter we observe by our five senses.

This will be attempted in the next chapter.

10 CAN OTHER UNIVERSES EXIST?

10.1 WHY ASK THE QUESTION?

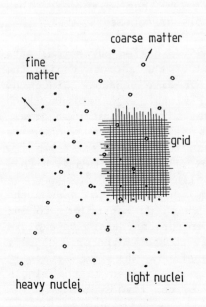

coarse matter

fine matter

grid

heavy nuclei

light nuclei

FIG.5 INTERPENETRATING MATTER ON ONE GRID

A picture has been built up in three spatial dimensions plus time and energy to explain our universe. The energy has three forms, positive and negative physical energy and the abstract psychic form needed for control. No other universes in other dimensions have been required. This does not mean, however, that other universes cannot exist. Indeed, the evidence considered in the previous chapter seemed to suggest that all observed phenomena cannot be fully explained by the Grid concept alone. For example, healers claim to have guides which to most people are totally invisible and undetectable in any way. Poltergeist activity also indicates the possibility that some interpenetrating system of matter may exist. Can our extended physics probe into this domain?

There is plenty of room for the elementary particles of other forms of matter to hide within our atoms. There is no need to postulate other dimensions in which the matter forms of other universes could reside.

This suggestion has caused some critics to complain that a contradictory statement has just been made. Any other forms of matter to which we are unable to couple must by definition exist in higher dimensions they postulate. Now I have a hologram. It is a picture of a spider about three inches long. It is a real image which stands out from a glass plate. This plate contains the interference pattern which causes the image to arise of course. I can pass my fingers through this image or I can measure its height above the plate with a ruler. I can in fact determine all its dimensions. Because I can interpenetrate the image without feeling any force I do not interpret this to mean that the spider exists in a set of higher dimensions!

Other kinds of matter can interpenetrate ours in a very similar way. If some special material from our universe could be made with identical atomic spacing to theirs, it might be possible to form an interface between the two systems, a kind of hard facing to prevent interpenetration. Then we would be able to see and touch the interface representing matter normally invisible and which ordinarily interpenetrates ours. Indeed such a mechanism might be involved in the sightings of ghosts or other materialisations.

In FIG.5 the Grid is represented by a set of horizontal and vertical lines. Crossing it is shown the cubic lattice of a certain kind of matter organised by wave functions emanating from the Grid. Then crossing in the other direction a crystal from a finer form of matter, also organised by the same Grid, is shown. At one place the two kinds of matter pass through one another without any kind of mutual interference.

From the viewpoint of physics other kinds of atom could be built having elementary particles structured like ours from positive and negative forms of energy, bound by the mediators of space. But the mediators need to be selectively coupled so that there is no interference between different matter-systems. Selective coupling is in any case required to structure a single matter-system so that electrons do not feel the strong force confined within the atomic nucleus and to produce the effect of electrical charge. Frequency, or alternatively wavelength, has been proposed with particles tuned in to mediators to provide such selectivity.

Michael Roll in television broadcasts has adopted this idea. He likens different systems of matter to the different carrier waves of the various television channels. They all exist together in the same space but it is only possible to tune in to any one of them at once. The others are still present but are not perceived. The idea that matter is composed of vibrations seems to have been around for many decades. Only matter-systems of a given waveband are made from waves, which are able to couple. The remainder interpenetrate and remain unsensed. In this way several interpenetrating matter systems could exist simultaneously in the same place. Each would be part of a whole unsensed universe, not existing in some remote region away from us but right here passing through us.

With the new insight gleaned from a close look at wave-particle duality it would seem that these matter waves are abstract. They correspond with the psychic energies we have been discussing. They are the wave-functions patterned on the Grid and the sub-atomic particles themselves are particles, not waves. These are required, however, to be able to recognise abstract waves within a certain frequency band and reject all others.

This idea suffers from frequency changes due to Doppler shifting, unfortunately. The "Doppler effect" can be observed when a train is heard with its whistle blowing. As it approaches the pitch seems high and this falls to a lower value as the train speeds past. The effect is due to the speed of the train causing a shortening of wavelength as it approaches and a lengthening as it recedes.

An additional "relativity" effect will also apply at speeds close to that of light. Einstein's theory of special relativity showed that a time dilation effect will arise when particles move at high speeds. In the new theory the same effect is predicted. This time it is due to the increased inertia of accelerated matter consequent upon the addition of kinetic energy. Any mechanical vibrations are slowed by the increase of mass as energy increases.

So that with the combined Doppler and relativity effects detuning would occur if high speed particles used only unprocessed frequency to govern coupling. This cannot, therefore, be the complete answer. However, the Grid possesses vast computing power and is able to sense the speed and position of particles. Therefore the necessary corrections can be made as a computer function.

There must be some other discriminator at work for creating the effect of electric charge. Wave frequency can restrict coupling to a given waveband; then an additional factor is needed to cut out all but selected kinds of mediator so that like charges repel and unlike attract.

Fortunately there is another means of coupling which can supplement frequency selection. "Spin coupling" is known and is able to provide the required supplementary effect. In our matter photons couple with electrons to produce the effect of electric charge. The virtual photons of space are therefore recognised in physics to be the mediators responsible. This will be considered in more detail a little later.

Photon-electron coupling is also responsible for vision. This is because vision depends on a process rather like the photoelectric effect taking place in the retina of the eye.

Hence, if a companion planet housed life-forms, operating on a different waveband, they would be

charged spheres i.e:-
they emit virtual photons ⎓⎓

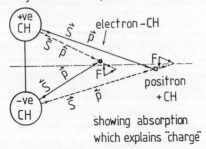

Spin rotation S   Momentum p

$\vec{S}$ ——)·→ $\vec{p}$ = positive

$\bar{S}$ ——(·→ $\bar{p}$ = negative

<u>FIG.6</u>   SPIN COUPLING SCHEME
can explain electric forces

invisible to us and could walk right through our walls without any resisting force whatever being felt. Yet in their world their matter would seem as solid to them as ours does to us. But they would appear like ghosts to us, like phantoms devoid of substance. But we could walk through their walls just as easily, so we would look like ghosts to them.

An illustration of the way a spin-coupling scheme can explain the electric force according to the new theory is given in FIG.6. Mediators of both positive and negative energy are continuously thrown off to create the effects of both positive and negative charge. But the negative charge throws off its mediators with spins opposite that of the positive charge. In this way when two negative charges interact, they absorb the positive energy mediators from one another whilst ignoring those of negative energy. When opposite charges interact the mediators of negative momentum are recognised so that attraction occurs. In this case the mediators are made of negative energy. It was shown in Chapter 5 that mediator absorption would lead to charged particles continuously gaining energy when in orbit so that a buoyancy model had to be used instead. The spin-coupling scheme is just as valid, however, but needs a different mechanism. In this case recognition of an incoming spin direction means that the charged particle offers a larger volume to that mediator than it does to one which is not recognised. Buoyancy-type forces are proportional to the effective volumes of sub-atomic particles multiplied by the partial pressure gradient of the mediators.

In order to account for the spin and magnetic moment of the electron, a study given in Chapter T.S.3 showed this could not be an elementary particle as is normally assumed. It needed to be a composite of at least two bodies of positive energy, coupled by mediators of the negative kind. Other sub-atomic particles such as protons and neutrons also need to consist of both forms of energy with one kind dominant, since they are also compound structures.

Hence the universe needs to be a composite system of positive and negative energies in more than one sense. It has already been shown in Chapter 5 that space without matter consists of an exact balance of positive and negative energies, which are mostly accounted for by the mediating particles on which the forces of nature depend. With matter added in concentrated lumps of net positive energy a tenuous halo of balancing negative energy needs to be superimposed on space. This halo stretches out for about a billion light-years from each object and causes the primary force of gravitation. Now each lump of matter is seen to be itself a composite of positive and negative with the positive form dominant.

10.2 OTHER POSSIBLE KINDS OF MATTER

From the foregoing it is clear that it could easily have been the negative energy component of matter which was in excess.

If the net mass of a companion system was negative, then all the forces of action and reaction required to produce motion would be reversed. The mass sign of all mediators also needs to be reversed. A detailed study, described in Chapter 5 and summarised in the TECHNICAL SUPPLEMENT (i.e. T.S.), has shown that such a matter-system would have exactly the same responses as our own.

This symmetry is illustrated in FIG.4[*] and also in FIG.14[**] (T.S.). So a negative-state universe could exist which would appear identical in every way to our own, assuming ours is positive. By observation it would not be possible to tell that the energy of construction was opposite from our own unless matter from the two could be made to interact.

But if it did exist, unable to couple with ours, then it would not be possible to tell the difference. If its physical energy exactly balanced ours, then space would also need to be in exact balance. Then there would be no mediators for gravitation and so no gravitational forces could exist. It follows that if other universes exist, they need to have predominantly the same sign of mass/energy as our own.

A second variant could exist in which the real sub-atomic particles were of opposite mass/energy. For example, the nuclei could be positive and have electrons of negative mass around them. A particularly simple coupling scheme then becomes feasible. The effect of like charges repelling and unlike attracting, as in our system, can then be produced by each object sending out mediators of its own kind of mass/energy. The end result would, however, be a universe acting much the same as our own. Hence from our viewpoint it would not appear remarkable.

But a third variant could exist whose behaviour would look most magical. It is a special hybrid case in which some atoms could have net negative mass whilst others were positive.
If these were exactly balanced, then the net energy of such matter would be zero. It would have no net inertia. Any life-forms constructed from it would be similarly balanced with no inertia. Then the most extraordinary properties can be predicted by physics. Firstly, it would be possible to accelerate almost to the speed of light by application of the slightest propulsive force. It would not be possible to kill anyone by firing bullets at them. The target would simply jump away. But even stranger properties can be

* See Page 86   ** See Page 282

predicted.

If two equal positive masses are free to move but are joined by a stretched spring which is released, then they will spring together and bounce apart. If now the experiment is conducted with one of the masses made negative, the same force will produce the opposite accelerating response on that element only. If the reader is unable to comprehend the next development, then reference to the T.S. is recommended, to the section explaining Newtonian mechanics using FIG.14 (T.S.).

The required extension to cover negative energy is also described there, referring also to FIG.4 in detail. An extraordinary phenomenon predicted by the extended Newtonian physics will then appear. Both masses will continuously accelerate away in the same direction! They will maintain their separating distance, so the spring force will not diminish. So they will gain speed indefinitely. Each member of the "accelerating pair" will gain in its own kind of energy and momentum, yet no conservation laws will be violated. This is because the values cancel so that the pair as a whole always has zero net energy and momentum. The response can be varied by adjusting the spring force "F".

The systems are compared in FIG.7. The masses could be atoms coupled by adjustable electric forces and built into living creatures. They would have the ability to adjust their weight at will to any comfort level simply by pointing an adjustable fraction of all the accelerating pairs downwards. By pointing some of them upwards to cancel the acceleration of gravitation they could levitate and live like a fish without obvious support. Then by pointing some in any desired direction they could accelerate within seconds to speeds close to that of light without feeling any force.

Even though the net inertia of the pair is always zero, change of speed cannot be instantaneous. This is because each component is made from a finite quantity of energy and so requires a force of acceleration. This force cannot be infinite. However, if of the order of

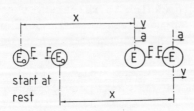

FIG.7 AN ACCELERATING PAIR

After acceleration "a" from rest to speed "v", energy "F.x" is added to rest energy "E_0" to give the total energy "E". i.e:- $E = E_0 + F.x$

But for the negative member:-

$$-E = -E_0 - F.x$$

So the net energy is always zero!

the electric force coupling atoms, this force can be very large as compared with those we normally experience. Hence accelerations could appear extremely high in comparison with those available to us. The pair would still be limited to speeds less than light. However, it has already been shown in Chapter 9 that an alternative "apport" system of transport is allowable by extended physics, which would permit change of location at a speed many times faster than light.

People in such a universe of balanced energy would have no energy problem. Any power requirement could be obtained by engines based on accelerating pairs. In effect they would be using pure creation as a source of energy. To scientists and engineers who have studied thermodynamics this statement must look like pure heresy. The First Law of Thermodynamics states: "Energy can neither be created nor destroyed," full stop.

All experimental checks made so far appear to support this statement absolutely. But the statement is based upon a physics which denies the existence of negative energy states. As soon as these are allowed, a different situation appears and it becomes necessary to relax the law to a substantial degree. It should now be written:-

"The net physical energy of the entire universe can never be altered and must remain zero at all times."

This permits the creation and destruction of physical energies, provided both kinds arise or disappear simultaneously in exactly equal amounts. The existing first law now appears as a special case arising when there is no possibility of change in negative energies. Furthermore, it will be shown later that with a carefully designed experiment a discrepancy in the accepted definition is predicted to arise if measurements are made with sufficient resolution.

Hence there is no reason why energy cannot be created from nothing. The only stipulation is that the conservation laws of energy and momentum should not be violated. With balanced systems of matter this is readily achieved because both positive and negative energies are simultaneously created in equal amounts. It cannot be regarded as impossible simply because we have not so far achieved such an energy-generating system. After all, the consensus of physicists is that the whole universe suddenly created itself from nothing. If everything arose this way in the first place, then why should it not be possible to achieve the same result on a smaller scale?

The internal engines of life-forms of exactly balanced energies could operate by adaptations of accelerating pairs and so companion people would never need to eat.

But their matter could have no mediators of gravitation emanating from it to produce a gravitational force. This is because their matter-system is already balanced. It was shown in CHAPTER 5 that our matter has unbalanced energy, so that the balance was achieved by a tenuous halo of negative energy stretching out for about a billion light-years to provide the primary cause of gravitation. So a companion universe of balanced energies would have no gravity of its own to hold its planets together. It would exist in the space compressed by our gravity, however. It would therefore be gravitationally coupled to our matter, even though uncoupled with respect to electric or nuclear forces.

In the TECHNICAL SUPPLEMENT it is shown that particles of both positive and negative physical energies will respond by accelerating in the same direction in space whose density is not uniform. They are both directed toward the places where the virtual particles are more closely packed, which in turn is caused by the presence of ponderous matter. Hence matter of exactly balanced energies will still fall just like our own. On hitting the ground, however, it will stop without any force being felt and there will be no natural weight. This is because the components are being pulled equally in opposite directions. If a living organism wishes to experience weight, then this needs to be deliberately simulated. Zero weight implies that accelerating pairs are not activated. Activation could be readily controlled by variation of the pair-coupling force "F" and by the alignment of pairs in a given direction. Pointing them all downwards would simulate the effect of weight.

It would be relatively easy for life-forms to point their pairs upward and control "F" to such a value as to exactly neutralise the natural acceleration of gravity. Then the organism would exhibit no weight and could readily levitate and hover.

Planets of matter in exactly balanced energies could not hold together by themselves and so they could not have an independent existence. They would have to use gravity "borrowed" from a companion planet like ours, and so could only condense about the planets of our system of matter.

At this point the reader may have noticed that the descriptions given for the behaviour of creatures made from exactly balanced energies closely match those of so-called "spirits". Hence the available data seems to support their existence in balanced energy states. This may seem an outrageous statement. However, reports of the kind given by Arthur Findlay, who studied their behaviour via direct voice mediumship, give descriptions which match the predictions just made to a remarkable degree. It is therefore most unlikely that the planets of other systems of matter will be found moving independently of the objects in our universe.

In the manner just described life could exist around any planet of our solar system. It could even exist on the Sun! This is possible because the absence of coupling between matter-systems would prevent the transmission of heat. It is also perfectly feasible that life of the kind we know and that of companion substance could exist on other stars and their planetary systems.

We need to look out for evidence supporting life on other planets of our solar system, but made from companion systems of matter. If they are found, this would confirm the idea of their matter being made from exactly balanced energies. It is possible that the primary purpose of our system of matter is simply to provide the anchorage for other forms. By chance it just so happens that our planet is suitable to support the life-forms with which we are familiar.

10.3   LIFE FORMS AS COMPOSITE SYSTEMS

It would be possible for living creatures, like us, to be composites, existing in several universes at once. All components would be controlled by the Grid to remain locked on to one another at all times. If each component were conscious at the same time, great confusion would result. Therefore it would be necessary to so programme the assembly that only one would be awake at any one time. The others would lie dormant, yet would assimilate all experience felt by their companions by storage in their collective subconscious memory-bank. When one component reached the end of its life, then another would awake and take over the business of acquiring experience.

An alternative explanation is that there is only one mind for each person. This would be patterned in the switching-system of the Grid itself. It would be surrounded by a filter/barrier to cordon it off from the

rest of the universe. In this way it would be protected
from overload whilst permitting adequate coupling to
relevant information. It would be programmed to act
through the matter-interfacing system of the lowest
order of the complete assembly. This means that our
physical brains would then provide the interface which
enables the body to be controlled. All other companion
matter-systems of the assembly would be bypassed in this
arrangement. They would lie dormant. Then the next
higher interfacing system would be activated at the
death-signal from the lower. In this case the
programming would ensure the single mind entity
interfaced through the brain of only a single system of
matter at any one time in order to avoid confusion. It
could be that schizophrenia results when the system is
imperfect and allows more than one interface to act
simultaneously.

Whichever explanation applies, it is clear that
several states of awareness could be potentially
available. Normally most could be filtered out so that
only one is perceived.

It is just possible that a weak communication link
between such states of awareness could exist. It is
therefore no longer reasonable, on the basis of what is
allowable by extended physics, to discount the
possibility of mediumship being a genuine gift. People
made in another system of matter might be able to coat
themselves in a special substance to prevent
interpenetration. Then they could speak to us directly
by causing pressure waves to arise in our atmosphere.
The interface could then also interact with our photons
to render the surface visible. If the coating was only
very thin, then the effect would be that of an
apparition. It is no longer reasonable, therefore, to
discredit people who claim to have seen ghosts walk
through walls.

This is not the only plausible explanation for
ghosts. A companion person might be able to transmit an
image telepathically via the Grid. This would affect
the brain in exactly the same way as input from neurones

coupled to the eyes. The person would see the ghost image superimposed on the image received from the eyes. It would not be a "real" image in the accepted sense but it would describe a real person made from interpenetrating matter.

Indeed both phenomena just described appear to relate closely to those which have come to us via the route of mediumship. Other supporting evidence arises from frequently reported "out of body" experiences. Some report accelerations to immense speeds without effort. All report hovering in space, looking down on their bodies. One woman reported being shot out from the Earth to look down upon the entire globe from a point far out in space. Then she described the return. At no time were any forces experienced. This could only happen for a body built from balanced energies. Of course the report could have been a fabrication. But it cannot be discounted on the grounds of contravening the laws of physics.

None of these reports can be dismissed as absurd on the grounds that they do not accord with normal experience. As already explained, they all accord with expectations for the physics of exactly balanced systems of energy.

Another most interesting point arises. No reports seem to describe the "ethereal" body being observed from the owner's Earth body. Unless reports appear which contradict this statement, then the evidence suggests that the mind is separate from our bodies. In this case the second hypothesis is supported. The brain needs to be thought of as simply an interfacing system, arranged to enable the mind to act through it for control of the Earth body.

It therefore follows that the basis of most religions concerning the possibility of humans having souls is supportable by extended physics. Creatures could exist in more than one universe at once, linked only by common gravity and the all-pervading Grid. The mind of an individual seems most likely to be a sectioned-off portion of the Grid, fitted out with a barrier to restrict the uncontrolled flood of information from the rest of the universe. If so, it could, from the viewpoint of extended physics, be clothed in several bodies of interpenetrating matter. One by one these could die off until finally only the mind-component remained. It would then be a disembodied mind. There is no reason why such a mind could not still interact with other minds, both disembodied or otherwise. It would live in a manner very similar to the computer simulation called "Life" originated by Conway.

Conway set up some simple logical rules to produce patterns of rectangular elements on a computer screen. Bordering elements would either create another next to it or die, with a final alternative of staying put, depending upon how many neighbouring elements existed. With these simple rules a whole unpredictable universe of effects arose. The game had no winners or losers, yet many people became quite "hooked". An ultimate life-form having a similar, but of course vastly more complex, basis is not impossible. After a few million years of such an existence, however, one imagines that things might start to get a little boring! It might be necessary to start again at square one just to retain sanity.

We may now be homing in on the reason why universes were created in the first place. They are deliberately isolated environments arranged so that the consciousness needs to work in order to ensure the survival of its temporary housing. Each matter-system is a deliberately contrived illusion.

The one to which we relate at the moment is, however, in dire trouble. The system has become unstable and is running out of control toward what looks perilously like a major disaster scenario. Let us therefore turn our attention to practical matters and see if our new insights can help us home in on a solution to our difficulties.

11 FUTURE PROJECTIONS

11.1 THE GLOBAL PROBLEM

Not "predictions". Projections can help us assess what is likely to happen if the course of the recent past is continued into the future. Then if the result is unsatisfactory, corrective action can be taken to produce a more acceptable future. The reason this chapter has been included is that the new theory offers new possibilities which need to be explored. It is particularly important at this time that any possibility, no matter how remote, be thoroughly investigated if it could help to save our planet.

The global population is now so high that the life-support system is seriously overloaded. Yet numbers are still increasing fast. The overloading is all too obvious. In a desperate scramble to grow more food to feed starving millions in the so-called "Third World" forests are being destroyed at an alarming rate to create more arable land. And after about three years it decays to desert. As the population continues to grow and requires an ever greater food-supply, so the resource base declines as land is laid waste. Already millions have starved to death. Just think what is likely to happen as this scenario continues to develop, with a projected doubling of the population by 2030. What can be done to save the situation? That is the pressing question.

There is only one answer but it requires the simultaneous implementation of two quite different strategies. The effective resource base needs to be

189

increased whilst at the same time the birth-rate is drastically reduced. The latter means that, overall, the birth-rate must be brought down to a level where it does not exceed the death rate within a time-scale of about twenty years.   Both strategies need to be implemented simultaneously. The effective resource base can be increased in two ways. One way is to provide more economical equipment and to exercise greater care in use. But there is a limit to such measures and they cannot be anywhere near adequate on their own.   The other way is to increase the actual resource base.

The combined increase of the effective resource base on its own will, however, only buy a few more years. Then catastrophe will be inevitable and worse than if this step had not been taken, unless the population issue is not simultaneously resolved.

It is more complex than this, of course. Another consequence of overloading is the increase of pollution of land, lakes, rivers and coastal seas. At the same time mineral resources are being depleted. The burning of fossil fuels and forests is creating the much-feared "greenhouse effect" by releasing carbon dioxide to the atmosphere. This is likely to raise sea levels and have adverse effects upon the climate. And then, to cap it all, holes have appeared in the ozone layer at the antipodes.   This means that ozone has been thinned worldwide. It will have an adverse effect on the growth of all plant life due to increased ultraviolet radiation from the Sun being allowed to reach the surface of the Earth. Unless the damage can be corrected, the food resource base will be further reduced just at a time when it needs to be increased.

It is well worth while taking an old book off the shelves; then, after dusting, it is to be opened at page 124. I refer to "The Limits to Growth" by the Meadows Group(118), published in 1972. They produced a most comprehensive computer programme to model dynamics of the global system. It took account of mineral and fossil fuel resources, population growth, food and other resources. It took into account a wide variety of other factors such as the greenhouse effect, pollution and the reduction of agricultural land area.

Page 120 shows FIG.35 giving a summary of the projected effects. Food per capita is shown to increase as farming methods are advanced until about the year 2100. Then this starts to plummet down again. The pollution levels go on increasing until about 2030 and the population goes on increasing until 2050. Then it starts to fall rapidly as life-spans are drastically cut by mass starvation. What is very worrying is that the trends it projected two decades ago are being followed to an uncanny degree of accuracy.

The book made considerable impact when it first appeared but unfortunately this was soon blunted by powerful lobbies. To big business it was anathema because they wanted growth regardless of any long-term consequence. They successfully debunked the projections by pointing out that the mineral resource base had not been adequately assessed. The authors admitted this in the book but they had done their best. To allow for possible underestimates they had then arbitrarily doubled the resource base. FIG.36 on page 127 shows this actually makes matters worse! The population peaks ten years earlier. The reason is that pollution effects run away and make the major impact by retarding the growth of plants. Meadows show the effects of other assumptions but in all cases total disaster is projected unless effective population controls are implemented.

This book should never have been debunked. Its projections show the real dangers we have to face. It will need a massive drive of the kind only normally met with in a major war situation, but with all the nations of the Earth in total cooperation if life on our planet is not to be almost extinguished. Even today few people are fully aware of the daunting magnitude of the task awaiting.

No solution is possible without tackling population growth, but this factor will be left until later. Let us first see how the new theory might help with the "Technical Fix" for increasing the resource base as part of a possible solution.

At present most of our system of energy supply used to give us electricity, power our cars and ships or fly our aircraft derives from fossil fuels. These are coal, crude oil and natural gas. When burned, carbon dioxide is created and is the major cause of the undesirable greenhouse effect. In addition the oxides of nitrogen and sulphur are emitted. These cause the acid rain which falls on forests and arable land. It has virtually destroyed the Black Forest and many trees in Britain already show signs of stress. The acids cause aluminium to wash out of soils into rivers and lakes to poison the fish.

A slow start has been made to remove sulphur dioxide from power-station exhaust-stacks. This uses limestone, which is converted to calcium sulphate. Apart from the high capital cost of the equipment needed, which is expected to increase costs by ten per cent, these measures will make severe inroads on limestone deposits. Many have declared this unacceptable.

Nuclear power has been proposed as a solution but this strikes a Faustian bargain. It is too dangerous as well as being too expensive. And the radiation hazard has been underestimated. The incidence of leukaemia around nuclear processing plant is evidence enough. Nuclear accidents have occurred, causing damage to ecosystems over hundreds of square miles. About six nuclear submarines have sunk, carrying their radioactive piles to pollute the sea-floor. It is becoming ever more clear that this solution is simply not acceptable.

To find a better energy source enormous efforts have been made, supported by multi-billion dollar budgets, to find a way of providing power by nuclear fusion. At present nuclear power is based on nuclear fission. The atoms of heavy unstable elements like uranium-235 split into radioactive fission products with the release of vast quantities of energy. Nuclear fusion, on the other hand produces much more energy by starting with light atoms such as deuterium, a heavy isotope of hydrogen, and fusing them together to build up heavier elements such as helium. This is claimed to be potentially more benign than the nuclear fission.

Fusion has proved much more difficult to achieve than was originally expected and physicists have stated that power from this source is at least forty years away. But we do not have forty years! Many advocate sustainable alternatives, such as wind, wave and solar power. These could help but are not adequate by themselves. Wind-systems have considerable visual impact on the environment. A farm of about 1,000 wind-turbines are needed to equal a single nuclear reactor, for example, and all this power needs to be backed up by fossil-fired stations. Hence their capital cost is an additional item. These so-called "free" sources of energy always turn out to be very expensive owing to high capital costs when any realistic economic evaluation is carried out. Even when account is taken of all reasonable attempts at economising it is clear that some other kind of solution is desperately needed. One promising resource which has not yet been tapped is the deep ocean.

11.2 ENERGY AND FISH FARMS IN OCEANIC GYRES

The land is already overstretched for food-growing and so it seems strange that so far the oceans have received little attention. Of the total surface of the Earth 70% is covered by the oceans and for the most part they are the equivalent of barren deserts. They support little life in tropical and sub-tropical regions owing to lack of nutrients in the photic zone. This is the top 100-metre layer able to support plant growth by light penetrating from the Sun. The nutrients lacking are the phosphates, nitrates and trace elements needed to structure the phytoplankton, the unicellular marine plants which form the base of the food chain leading to harvestable fish.

With these factors in mind Eric Marsden and I did our best to stir up the enthusiasm needed to start some experiments in the deep ocean. Reference was made in the opening chapter to an exhibition we set up called

"ENERGY FARMS IN OCEANIC GYRES". This was actually the second. The previous year our "STRATEGY FOR SURVIVAL" held at Olympia in 1977 had also introduced the topic. Marine farming of the deep oceans has to depend upon the artificial upwelling of nutrient rich water from the deeps. This has already been proved to work.

Georges Claude, having made a considerable fortune with the development of commercial fractional distillation methods for the bulk manufacture of gaseous and liquid oxygen and other gaseous components of air, carried out experiments in Matanzas Bay, Cuba. His aim was to harness vertical temperature differences in the tropical ocean, of around 24 degrees Celsius, for the generation of power. An "upwelling pipe" brought up cold water from the deeps for use as coolant. The warm surface water could be made to evaporate at sub-atmospheric pressure and the steam so evolved could be used to drive a turbine for power generation. The steam then condensed in contact with the upwelled cold water. The system worked but the project was beset by corrosion problems. The project itself was a fiasco and ended with Claude dumping the whole mass of equipment into the ocean.

Ironically, the real success was from spin-off. The local fishermen were suddenly delighted to find their catches had increased threefold for a time. It was later discovered that the upwelling process was responsible. It had produced accidental fertilization of the photic zone!

Unfortunately carbon dioxide is brought up as well as other essential nutrients, but in amounts relative to plant uptake, of more than six times as much as can be utilised. Using upwelling from depths of 600 metres or so would therefore exacerbate the greenhouse problem. To provide a reversal of the greenhouse effect phosphates need to be leached from the bottom sediments, where their concentration is about 1,000 times that in the water above.

In the new project floating seaweed was the target crop, from which energy could be provided as a dried

flake. Sargassum Fluitans and Sargassum Natans already
float around in that area reproducing vegetatively
whilst confined by the North Atlantic gyre. Hence the
proposal was to increase its productivity. Later ideas
showed that substitute crude oil would be a better
product made on site by heating wet seaweed at great
depth. This released carbon dioxide there and locked it
up in seawater more effectively. Such oil could also
provide all the needs of the petrochemical industry for
the manufacture of plastics. A second product would be
fish, since a computer simulation of the marine
ecosystem, specially written to study the effects of
upwelling, showed that about half the nutrient supply
would be taken up by phytoplankton.

With fuel and fish produced together, economic
viability was forecast despite the 4,000-nautical-mile
distances involved, because fish would be transported on
the fuel product carriers. In addition, instead of these
carriers returning empty they would carry sewage sludge.
This is normally dumped in the North Sea and creates the
over-fertilisation which leads to toxic algal blooms.
Dumping in the North Sea, therefore, sludge can only be
regarded as a serious pollutant. Spread over greater
areas of the deep ocean, it would be recycled to
positive advantage at very low additional cost. Hence
marine energy/fish farms in the open ocean could provide
a unique combination of advantages. They would improve
the carrying capacity of the planet.

Other advantages would be the revitalisation of the
shipbuilding and steel industries. These would absorb
the workforce released from consumer product industries,
as these are reduced to comply with the zero consumer
growth scenario at which the world needs to aim.

The North Sea is already badly overfished and it is
essential that new resources are made available.
Unfortunately we failed to attract the necessary
interest and all our efforts achieved was a relatively
trivial project aimed at controlling fresh water weeds
in the Kawartha lakes of Canada. It was a tragedy,
really, because at the time 24,000,000 deadweight tons

of redundant tankers were laid up around the world.
Nobody had the slightest idea what to do with them, so
they just lay rusting at their moorings. They would
have provided a cheap start for marine farms.

Marine farms in the deep ocean would still be worth
considering for the provision of protein as fish for the
rising population and for the raw material of plastics.
But to provide all the fuel needs of the world the
entire gyres of both the North and South Atlantic oceans
would need to be cultivated. It would be a giant task
and would take a calculated fifty years to complete. No
other project has been considered which would be able to
lock away more carbon dioxide than is produced by
burning of the fuel it creates. Indeed it seems to be
fundamentally impossible that any other could have such
an advantage. The sheer scale of operations and the
reluctance of leaders to face reality make it unlikely
that the solution just outlined will ever be tried.

It is worth searching for other alternatives.
Indeed, so great is the problem that no stone should be
left unturned. Even the most remote possibility of
finding a new wholly benign alternative energy source
needs to be explored.

11.3 A NEW ENERGY SOURCE?

A considerable body of literature is already in
existence concerning negative mass and negative energy
states, as described in the references (301) to (314).
They were first shown to be allowable in physics by Paul
Dirac about 1930. Robert Forward(307) has used a
similar idea to the accelerating pair described in the
last chapter to show that if a supply of negative mass
could be found, then the perfect space-drive could be
constructed. Because the discovery of useable matter of
the negative kind is very unlikely, he only suggests the
idea can have value for use in science-fiction stories.
The same idea for space-drive is discussed by Alan Guth
in *The New Physics*(108). Both he and Forward, however,
rely on a gravitational coupling between the two kinds
of matter.

From investigations of this kind it always transpires that the pure creation of energy in equal and opposite amounts needs to be regarded as permissible. This means that a very dramatic reversal of previously held concepts is starting to become acceptable to physicists. The old definition of the First Law of Thermodynamics has already been undermined and needs revision. A suitable new definition was given in the previous chapter. The meaning is very clear. It is feasible, in principle at least, that power could be generated by pure creation.

The present study developed during 1982 in a way paralleling the ideas of R. Forward. It was necessary to postulate the existence of negative energy states to explain creation in the "Big Bang". First ideas suggested that two universes had to arise simultaneously, one from positive and the other from negative energy. It was then easy to show that logical inconsistencies would arise if cross-coupling by electric forces were present. Therefore the two forms of matter had to be mutually invisible and would interpenetrate one another. All efforts to publish this material in scientific journals met with total rejection and so a start was made to create a science- fiction story as the only means of expression. It may be of interest to summarise the plot.

John, the hero, thinks of the idea and by psychic means locates a parallel planet of negative matter interpenetrating our own. He finds it has people similar to us who also suffer from an energy problem. They join forces to devise a power-unit capable of generating both kinds of energy simultaneously and each uses their own kind. John then tells the BBC that he is going to astound them by demonstrating his new invention. But he tells them to have a camera ready on the rooftop since he intends to arrive in an unusual way. This he does sitting in an armchair clutching at a cable pointing vertically upward and attached to the seat of the chair. The cable seems to be attached to nothing at its upper end and the whole assembly is observed flying majestically through the air toward the camera crew. Of course the readers know that at the top is a chunk of invisible negative matter providing an exact balance so that levitation and acceleration in any direction can be achieved.

He causes a great stir and the entire world is both baffled and astonished. His engine, demonstrated in the studio, produces power without any fuel supply being used. Little do the viewers realise that an invisible part exists and that in the other universe the same thing is happening. There is an invisible man on the other end collaborating in the control of negative power at his end. John sets up a business making levitation transporters similar to his chair by collaborating secretly with the other side. They have a deal in which they make parts for one another. The story ends in complete mayhem as large balls of concrete are seen flying through the air and crash into cars, houses and people. Also people start falling to the ground as their invisible supporting negative masses hit objects made from their own system of matter. The invisibility of each to the other has created an insurmountable problem.

I never completed the story because my friends soon made me realise that I was no story-writer. A pity. This all happened when the present theory was in its embryo stage. The existence of a universe of negative matter is very unlikely because the balancing negative energy states are needed to provide the primary mediators for the gravitational force. So negative energies exist as part of space and also provide the components holding atoms and sub-atomic particles together. The quest for a practical system of power generation by pure creation needs to adopt a more subtle approach.

The new theory of quantum gravitation, however, has thrown up a new possibility for an energy source, which, if confirmed, might provide a complete solution to the difficulties of energy supply at low cost and without pollution. If history is followed, it will be seen that energy crises have happened several times before. Each time we have somehow been homed in upon a solution. It is possible that history is repeating itself. The possibility needs to be explored. We will take a close look at this spin-off from the theory.

In Newton's theory of gravitation a force acted on an object pulling it downward toward the centre of the Earth. In free fall this force caused it to speed up. It accelerated in a downward direction and added kinetic energy to the object equivalent to the work done by the force acting over the distance of fall. The same concept is retained in the extended version. Here, however, the force of gravity acting on an object is proportional to its total energy, defined as its rest energy plus any kinetic energy. That this has to be so is argued in the TECHNICAL SUPPLEMENT. Briefly, light is kinetic energy and since it is bent by gravity it is known to fall like matter. Hence the kinetic energy of any object will also experience the force of gravity. So the kinetic energy of an object will add to its weight.

The validity of this deduction is demonstrated by the success of the new theory, which not only fits the experimental checks just as well as Einstein's theory of general relativity, it relates the magnitude of the gravitational force to that of the electric force as well! Nobody has been able to do this starting from general relativity. Even Einstein did not manage this despite many years of effort. The superiority of the new approach is amply demonstrated by the comparison given in TABLE T 11. Hence it seems very reasonable to suggest that an object possessing kinetic energy, due to some kind of motion, will weigh more than it does when the motion is removed. Before going further it is necessary to make sure some basic principles of mechanics are understood.

Some confusion may arise because Einstein used an interpretation of free fall different from that of Newton. Hence this needs to be reviewed. Einstein said that a man falling off a roof would feel no

gravitational force. He deduced that no force could
therefore exist for an object in free fall. Instead it
moved along a "geodesic" in curved space-time. It is
true that a person would feel no force when falling,but
it does not follow that Newton's concept is wrong.

For example, eight balls can be imagined so placed
that one touches each corner of a cube. If both cube and
balls are dropped simultaneously in a vacuum, so that
the effect of air resistance is not present, then all
objects will experience equal acceleration. All the
balls will remain touching their corners as they all
fall together. Hence if the balls were fixed to their
corners, no force would exist across the fixing points
during free fall. Similarly a man would feel no force
between any parts of his body, because no force would be
transmitted from one body-cell to another. This
conclusion is reached even though an accelerating force
is present on all particles, because the force per unit
of mass is equal for all. Each atom has exactly the
same acceleration as any other.

The situation is different from the acceleration
imparted through, for example, the back of a car seat.
Here the accelerating force needs to be transmitted to
all atoms of the body by forces acting from atom to
atom. The resulting stresses can be detected by the
nervous system so that the accelerating force is
experienced. In free fall each atom is subjected to the
accelerating force from within itself and so no inter-
atomic force is produced.

It cannot be inferred, therefore, that because no
force is felt no accelerating force exists. Despite
Einstein's unquestioned genius he does not seem to have
appreciated this simple point. Hence the rival
conceptualisation needs study.

A simple thought experiment can be set up to see
whether in principle net energy could be produced by an
asymmetric gravitational cycle. By this is meant a
cycle in which a lifting process and a lowering process
are incorporated having different kinetic energy from
each other. All processes are to be considered ideal in

that no frictional effects are present. It does not concern us that this is a situation which cannot be achieved in practice because it is only the theoretical principle which is being explored. In the TECHNICAL SUPPLEMENT it is shown that the "rest energy" of any object remains constant whatever its level in a gravitational field. This is the energy from which the object is made.

Now let an asymmetric gravitational cycle be considered. An object is first slowly lifted on a cable to a great height. This is the first process of the cycle, and lifting work is done equal to the weight of the object multiplied by the height difference through which the weight was lifted. The weight is the force exerted by gravity and is proportional to the rest energy of the object. Then the object is dropped and accelerates in free fall. The force of gravity still acts in the new theory, even though it is no longer perceived by internal stresses set up within the object itself. This force, in causing acceleration, adds kinetic energy to the existing rest energy, so that the average total energy during the fall is greater than it was during the lift. Now the new theory is based on the force of gravity being proportional to the total energy. The fact that predictions made from it meet all experimental checks gives confidence that this deduction is true. It follows that the work done by gravity on the object in free fall exceeds that done in lifting slowly on a cable. So the final kinetic energy on hitting the ground will be slightly in excess of the energy supplied by work done during lifting.

This is in contradiction to the First Law of Thermodynamics, which implies that exact equality should obtain. Some will use this to suggest the theory has therefore been flawed but such is not the case. What it suggests instead is that "gravitational potential energy" is not a true energy form. It is a pseudo-energy form which is nevertheless useful in ordinary calculations. In simple mechanics the term "total energy" includes gravitational potential energy. The latter is the same as the work done in lifting the object from a reference position. This mechanical work is a form of energy and is considered to be stored "somehow" in the form of gravitational potential energy. But nobody knows just how this energy could be stored in space.

Then an object in free fall is said to have a total energy which remains constant. As it falls, potential energy reduces as kinetic energy increases, so that total energy by this old definition remains constant. As level reduces the stored gravitational potential energy is released.

In the extended physics the idea of energy being stored in space needs to be abandoned because it no longer fits the facts. Potential energy cannot vary with the speed of the object, as was shown earlier to be the case.   Therefore the concept must be wrong.   Now, however, negative energy states are allowed and this comes to the rescue.   No longer is it necessary to postulate that total energy is constant in a state of free fall in order to satisfy the First Law of Thermodynamics. In the new concept, total energy is the sum of rest and kinetic energies alone.   Then when a ball is thrown upward, its total energy continues to reduce until the zenith is reached and the object has only its rest energy. As it falls back again its total energy increases. The old First Law is violated during both ascent and descent but in the new definition given in the last chapter it is satisfied.

What is happening is that as the ball rises and positive energy is reducing, there is an interaction with the virtual energy of space responsible for the force of gravity. The negative energy of space reduces at the same rate as the positive energy of the object so that the net energy of the universe remains at its natural zero level.

So as an object rises, energy of both signs is being destroyed and as it falls again they are re-created. Hence in the new theory the destruction or creation of energy from nothing occurs quite easily.   What is difficult is to arrange for a continuous cycle able to create a net continuous output of energy of both kinds. Only the positive kind can be detected or utilised, whilst the negative kind is added to existing energies of space. The conclusion that energies can be created from nothing cannot be regarded as absurd simply because this has been accepted as an impossibility since the time of Newton. If it were indeed an impossibility, the universe could never have been created in the first place.

If it is true, then why has the effect not been observed? This is a reasonable question. The answer is that for the kind of gravitational cycle used by way of example the effect would be so small as to be totally undetectable. If a kilogramme is raised two kilometres and dropped, the gain would be only 10^{-13} joules as compared with the 19,620 joules used in lifting on the cable.

11.4   AN EXPERIMENT IN PURE CREATION

An experiment could, however, be designed to create an observable effect if positively charged atomic nuclei, called ions, are used together with strong electric fields. This addition makes it possible to greatly increase kinetic energy per unit mass of particle for the descent. The ion will be fired upwards at a low speed and after rising freely to a point near the top of the apparatus it is turned through 180 degrees by a local magnetic field, so that it points downwards. The force acts perpendicularly to the direction of motion in the magnetic field, whose lines of force are horizontal. In consequence the direction of motion can be reversed without change of speed. Reversal is achieved without any loss or gain of energy.

Then the ion is accelerated to a very high speed by a static electric field produced by fixed electrodes which are so arranged that the ion does not touch them. This adds kinetic energy so that weight is increased. Then near the bottom of the fall an adverse electric field is situated. The electrical potential change in

this adverse field is made exactly equal to the fall
previously arranged, so that on exit the electrical
potential of the ion is the same as it was at the top of
the trajectory. Hence all the kinetic energy imparted
near the top of the apparatus is recovered again near
the bottom. The weight during the falling phase is
thereby caused to be greater than that during the rise.

The net effect is a predicted gain in electrical
potential of the ion, for the complete cycle of rise and
fall. The ion has returned to its original position and
so could carry on indefinitely completing cycles and
gaining an increment of energy at each. The gain per
cycle would be difficult to measure, however. In an
apparatus ten metres high and with a negative potential
of a million volts used in the falling stage a gain of
only one-billionth of a volt can be expected.   The
amount is just about detectable. If the apparatus is
inverted, then a voltage loss, indicating energy
destruction, is predicted. The net gain "δV" volts for
a voltage equivalent "V_1" of upward ion speed and an
equivalent "V_2" for downward speed, over a vertical
height "y" metres, can be calculated from the
expression:-

$$\delta V \;=\; \frac{n.g.y.(V_2 \;-\; V_1)}{c^2}$$

Here g = the acceleration due to gravity = 9.81   m/s^2
     c = the speed of light = 300,000,000 m/s
and n = 1 for a singly-charged ion.

Electrons could be used instead of ions but for a
given voltage change they experience much higher
accelerations. Unfortunately energy is radiated away
from accelerating electric charge and so produces a loss
having a similar effect to friction. This loss equals
the output when the voltage change is a few thousand
volts.   Then as the voltage is increased the losses
become greater than the predicted output. On the other
hand, for protons and heavier ions the loss is

negligible. The complete cycle is then negligibly different from the ideal frictionless type.

If this experiment were carried out and the predicted result returned, an immense milestone in scientific progress would have been made. There are three aspects:-

Firstly, it would prove that gravitational potential energy is only a pseudo-form, that it does not really exist. The result would infer the existence of negative energy states because without these the First Law of Thermodynamics could only be assumed approximate.

Secondly, the difficulties arising from Tryon's (123) concept would be fully resolved. He assumed from the conventional definition of gravitational potential energy that this was a negative form of energy. Then he showed its magnitude was about equal to the rest energy of matter in the universe. He concluded that in consequence the universe could have arisen *ex-nihilo* or from nothing. The zero or datum point was taken at an infinite distance from the place where the Big Bang was centred and from this reference point the gravitational potential energy appears negative. But the datum has only an arbitrary location. It is just as valid to use any other place as reference. If the reference point is taken separately for each particle, at the place where it is created in the assumed "Big Bang", then this energy appears positive, not negative at all. So which is it? Our previous deduction answers the question. It is neither. It is not a real energy form at all.

It is a pseudo-energy form and in consequence it cannot be the factor making the net energy of the universe zero. Hence only negative energy states are left and the deduction provides further confirmation that these states just have to exist. It was this argument which Professor Vigier agreed was true by his letter quoted in "ACKNOWLEDGEMENTS".

Thirdly, the new theory would be supported if the experiment returned a positive result. But Einstein's theory of general relativity, based on curved space-time, would be shown invalid. Energy creation by an

asymmetric gravitational cycle is not predicted by
Einstein's theory, because his concept is based on the
idea of no force of gravity being experienced in free
fall. Hence no work is predicted to be done and so no
difference in potential is predicted for an asymmetric
gravitational cycle. Indeed, in Einstein's theory a
force is produced only if the acceleration differs from
that of gravity. This is why he had to introduce the
principle of "equivalence" between gravity and
acceleration. He considered a person in an accelerating
reference frame and showed that exactly the same effects
would be experienced as for a non-accelerating frame in
a gravitational field.

A number of experiments have been carried out to
check this principle. These are expertly described by
Clifford Will(124) and seem very convincing. The
experiments of Eötvös and Dicke, for example, used the
gravitational fields of the Earth and Sun respectively
and balanced them against the centripetal acceleration
of the Earth caused by spinning about its polar axis.
They both used a beam-balance suspended on a fine wire
so that it remained horizontal. On its ends were
attached objects made from very different substances, so
that it was known that their chemical binding energies
would differ. They would also have different ratios of
electrons to nucleons. Located at mid latitudes, the
rotation of the Earth produced a horizontal component of
centripetal acceleration which needed to be balanced by
a horizontal component of acceleration produced by
gravitation. If, then, the differing binding energies
had any effect, a torsional deflection would be
observed. None was found.

Another type of experiment was designed to test the
effect of differing gravitational binding energies.
Very precise measurements of the orbit of the Moon were
made to see if the difference between the gravitational
binding energies of Earth and Moon had an effect. None
was found.

Will says that these tests are totally convincing
and prove equivalence. Then he goes on to state that

only the concept of curved space-time is compatible with these results. Hence Einstein's concept is proven as far as he is concerned. This conclusion is supported by the consensus of theoreticians in the field of gravitation and so most people would accept its validity without further question.

Now the astute reader is asked to consider the alternative. The rest energy of an object is postulated to be that of its component sub-atomic particles plus any binding energies. The gravitational force is still proportional to total energy, being the sum of rest and kinetic energies. But inertia forces are also proportional to total energy based on this definition in the extended Newtonian. Hence all the equivalence checks are satisfied just as well as by Einstein's concept. Indeed, working from the experimental checks made to prove equivalence, the above deduction arises. It can then be used as the starting point of the extended Newtonian physics. Hence once again the two theories parallel one another.

It is indeed most dangerous and quite illogical to state that the equivalence checks show that gravitation can *only* be explained by the curvature of space-time in order to exclude alternatives.

Altogether, if a positive result were returned by the new experiment, a whole new vista of scientific understanding would open up. It would show that in principle man would have discovered the secret of pure creation! My personal rating for the chances of this experiment ultimately returning a positive result are very high. I would put it at 99%, because the new theory on which the result depends satisfies all the experimental checks without the difficulties which have prevented Einstein's general relativity from being integrated with quantum theory.

There is little prospect, however, of generating commercial power from asymmetric gravitational cycles. The amount possible is absurdly small. What the experiment would achieve is the proof that the pure creation of energy, positive and negative in equal amounts, is both theoretically possible and practically achievable. Once this has been confirmed, even on the most minute scale, then the door will be open for research into the possibility of commercial power generation by pure creation.

There remains one small hope for harnessing gravitation. So far only linear motion has been considered. Gyroscopic action has not been introduced. The effect of precessional motion which gyroscopes exhibit is readily explained in terms of Newtonian physics. A torque is generated at right angles to the direction in which the axis of spin is turning and mathematics predicts this exactly. The theory also shows that there should be no unexpected effect like anti-gravity. Some people say, however, that gyroscopes can produce a demonstrable anti-gravity effect despite negative theoretical predictions. Hence a question mark has arisen and clarification is needed. It could be that some new effect will be discovered which will enable energy to be harvested from asymmetric gravitational cycles, though this seems unlikely.

The greatest hope for power by pure creation seems, however, to be offered by non-gravitationally based cycles.

If I am asked to give an assessment of the likelihood of this being achievable, based on pure logic and nothing else, then I have to say the chances are small. This is because the result depends on another assumption having to be satisfied which will not be considered here. I would have to rate the chances at less than ten per cent.

But this is based on logic. My hunch is that the prospect is not only feasible but that this is the means by which all power will be generated in the future. The only justification for allowing the hunch to override logic is that my theoretical work has been partly driven by hunches. During the last few years, when struggling to derive a solution for quantum gravitation, I have followed one hunch after another. If my hunch told me to look for a solution in a certain way, then I looked

at it that way. In almost every case the hunch method
has ultimately paid off, though not without some blind
alleys being followed. I have therefore developed a
strong reliance on the hunch for indicating the
direction in which to think.

11.5 PROSPECTS FOR THE NEW SOURCE OF ENERGY

What would the world be like if this hunch paid off?
It is worth thinking about this because it might in fact
come to pass. In the first place power units would take
the form of vacuum tubes like those in the television
set. Electric power would be dragged straight from the
quantum vacuum without the use of any moving parts being
involved. Consequently energy from this source would be
obtained at low capital cost and there would be no need
for maintenance. Running costs would not exist because
no fuel supply would be required. And as a further
consequence there would be absolutely no pollution of
any kind. How could there be with no exhaust?

In this way the damage caused by acid rain would
gradually heal because this source of pollution would
entirely disappear. Also the greenhouse problem would
be greatly eased because power generation would not be
accompanied by the production of carbon dioxide. If at
the same time forest burning were stopped, then the
problem would be completely solved without developing
marine energy farms. Some would still be desirable,
however, as a source of raw materials for plastics and
for boosting the fish supply. Then all the fossil oil
and coal could be allowed to remain in the ground.
There it could remain, passively continuing to fulfil
its proper task of locking away carbon dioxide forever.

There would be no need for coal-miners to go down
dirty pits. No need for messy oil-wells and all the
marine pollution caused by spills. All the ugly
transmission lines would come down, because every home
and factory would possess its own power supply. There
would be no need for any power-stations so all could be
decommissioned, whether they be coal-, oil- or gas-fired
or nuclear plant. No petrol pumps would be required.

This is because all the cars would be powered by electric motors with energy supplied by vacuum tubes under the bonnet. Ships would be similarly powered. Aircraft would have infinite range. Their jet engine compressors or fans would be direct-driven from high-speed electric motors. The fire risk from accidents would no longer exist.

In short, there would be happy smiling faces all around! Except for the face of the odd vested interest which might just frown a bit.

These frowns would have to be given sympathetic consideration and eased by restricting the rate of change to levels which could be absorbed. Their problems might be eased by giving priority to making the new power systems available to the Third World. Their people would not then need to carry on using wood for fuel and so one major factor causing the present attrition of forests would be removed. This would slow down the changeover in the developed world and so ease the financial losses which might be incurred in any such major transition. Ultimately all people in the world would have cheap non-polluting energy on tap.

Let us not be carried away by these dreams of Utopia however. They are not impossibilities but the major global problem still remains unresolved. This is the problem of population explosion. Let us now turn to this aspect and see how the abstract part of our new theory can help.

11.6 POPULATION CONTROL

It may be easier to appreciate the global problem discussed earlier, when reference was made to the work of Meadows, by looking at a scenario which has already developed. About 40 years ago the population of Ethiopia was less than 10 million, it was 40% forest and agriculture could just about cope. But the average family size was eight, so the population was growing fast. It meant a fourfold increase every generation. They chopped down the trees to make more land available for growing food, but they did not understand the subtleties of the natural ecosystem.

Trees store water, then let it steadily evaporate to come down again as more rain.   So the rainfall is amplified beyond the level provided by the primary water carried by winds from the ocean.   There is also a "flywheel effect".   Rainfall tends to be spread out more over the seasons.   The water eventually finds its way to the sea in clear streams and rivers but as a steady trickle which does no harm.   But chop down the trees and the climate becomes more arid.   Any rain which does fall tends to come down in short sharp deluges.   Most of it runs straight away off the land and is lost to agriculture.   Then, to add insult to injury, it takes away the topsoil with it, so reducing fertility.   In totality the land is rendered less able to carry crops. Instead of food resources increasing, they diminish as the trees are felled beyond a certain point.   The Ethiopians created their own semi-desert through lack of knowledge.   New Scientist some years ago gave details of rainfall in Ethiopia over several decades.   A steady decline was superimposed on sharper ups and downs.   The decline correlated roughly with the loss of forested land.

The Ethiopian population carried on rising regardless of the diminution of their food supply and now they have 40 million on land able to carry less than it did when it was only 10 million.   We are all upset when we see television pictures of the starving children.   The instinctive reaction is to send them food.   Bob Geldof has raised considerable sums for aid, then next time crops failed and the same situation arose he was surprised.   He said the same situation should never have been allowed to occur again.   Now I admire his spirit and his desire to help.   This is genuine, there is no doubt of that.   But he shows no deep understanding of the situation.   It is **impossible** to prevent the same situation from arising again and again. Suppose we simply go on sending food to prevent starvation without strings attached.   Then in another forty years there will be 160 million Ethiopians on land

only capable of supporting ten million. They cannot
spread out across their borders because the neighbouring
land is already occupied and surrounding populations are
also growing.

If Ethiopia were a single isolated case, then its
difficulties could be resolved. Unfortunately this is
not so. Ethiopia should be regarded more as a foretaste
of things to come. The entire Third World is heading
the same way and numbers will soon be too large to
permit of solutions which depend upon food aid alone.
We are faced with a situation unprecedented in history
and virtually impossible to resolve!

The very act of kindness ultimately makes the
problem worse because there are then more people to
ravage the ecosystem. Ultimately the aid has to stop
and then the crash, measured in numbers of deaths from
starvation, will be greater than if no aid had been
given in the first place. This highlights the problem.
Unless it is properly understood in a mathematical way,
the very act of helping can make matters worse instead
of better. Aid needs to be accompanied by some form of
restraint so that it is not translated into more mouths
to feed at the same level of starvation.

It is no use trying to refute the above analysis by
pointing out that warfare has existed in Ethiopia for
many years. This has certainly caused a bad situation
to be aggravated, but the removal of war would not
eliminate the dilemma caused by population growth.
Indeed, as a struggle develops to grab ever more scarce
resources, more conflict situations will arise.   It
seems probable that this factor will be the normal
accompaniment of the developing problem, making it even
more difficult to resolve.

Some people have argued that all that needs to be
done is to provide aid on such a scale that population
growth would be out-accelerated. It is pointed out that
once upon a time growth rates were high in the developed
regions. They fell to the replacement level as soon as
living standards rose. Therefore raise the living
standards of the Third World to our level and, hey

presto, the problem goes away of its own accord. We
have already shown by our own past history that this is
a feasible proposition. Hence why not implement this
proven cure right away, it is asked.

This is the simplistic view of the non-
mathematically-minded, unfortunately. Mathematics needs
to be pumped in to quantify the problem. This has
already been done. A sequel to *The Limits to Growth* was
Mankind at the Turning Point by Mesarovic and
Pestel(119), which came out in 1974. They split the
world into a number of regions which interacted with one
another and in this way refined the computational
approach which Meadows had used. They explored the idea
of out-accelerating population growth and found it to be
totally impractical. To achieve this in just one
region, South East Asia, would have required a supply of
food equal to the entire output of the whole world.
Unless the supply exceeds a certain threshold level,
then the situation worsens at a far greater rate than if
no attempt at a solution had been attempted. This
solution is only possible when populations are still
relatively small. Beyond a certain size some form of
positive control is needed. Yet nobody wants to take
away freedom from the individual. The only possible
acceptable alternative is some kind of change in
attitude.

Attitudes and the sense of morality or knowing right
from wrong are ethical values. They are the province of
religious and political leaders. Our new theory can
help simply by showing that the faiths people have held
since antiquity are supportable by extended physics.
Faith no longer needs to be considered as the sole
supporting prop. Nature seems to fill all possible
niches. This is the experience of science, whether of
the biological or physical kind. And the new theory
shows there are vacancies for other universes and other
forms of life which we cannot sense and which
interpenetrate our own. It shows that life-forms could
be composites living in more than one universe at once.

The new theory offers convincing evidence that a

Creator was needed to form the universe in the first place. Indeed a Creator is needed all the time to continue to control the structure of matter. All sub-atomic particles are constantly regenerated, as each can only have a fleeting individual life and the place of reconstruction has to be determined by the Grid using superimposed wave functions as numbers. Since physics shows it to be possible that people could be composites, the concept of people having souls is supportable by extended Newtonian physics.

Without this support faith has had to be relied upon and this is very dangerous. Other faiths appear as threats and this has been the cause of endless violent conflict. Furthermore, once incorrect ideas have been implanted, such as that of the Earth being flat, then it is very difficult to make correction. Any change to a part can be seen to undermine confidence in the whole. Even if the leaders want to put things right, the followers will not let them. This is because ideas implanted in the subconscious at an early age cause any later ones, which are seen to be incompatible, to be rejected. The would-be reformer is likely to fall between two stools, feeling the wrath from both leaders above and the congregation below. There is in consequence a very strong incentive to suppress advancement and change in order to protect the power structures of established churches. The new theory can help to allow changes to be made without harm to these power structures. The power structure of a leading principle is not in itself a harmful thing because leadership is useful. Chaos might reign otherwise. A power structure can become harmful, however, if it perceives new advances as threats to itself. Its response is to try to destroy them. New knowledge which offers the lifeline of showing basic beliefs to be supportable can be accepted more readily, because it no longer poses a threat.

Now what may be right for a bygone age can be wrong for the present. Right and wrong cannot be regarded as absolutes correct for all ages. Some commandments at least need revision from time to time.

It was right 1,700 years ago to say "Go forth and multiply" because at the time the Earth was underpopulated. Human life could have died out at that stage. Now the same directive is tantamount to global suicide. Yet many religions still say "Go forth and multiply." They do not seem to dare to change anything in their creed, because they have faith and faith alone to hold them together. Surely when faith is supportable by physics these attitudes can be relaxed?

There is no way wave-particle duality can be satisfactorily explained using theories which do not support a creationist scenario. If this statement is doubted then chapters 2 and 3 should be read again. Previous explanations which try to avoid this conclusion simply do not work. They require impossible assumptions like infinite numbers of systems of matter existing simultaneously in one place, the act of simply looking at an object causing it to spring into being from an unresolved ghost-like wave-function, or other similar absurdities. It would be quite impossible to find a solution, based on common-sense, which does not support a creationist theory. Therefore no danger now exists that the important basic beliefs of any religion could be toppled by making necessary amendments. Religion need no longer fight against science as far as basics are concerned. These beliefs are the existence of the soul, a Creator and an ultimate purpose for life. All other factors are mere window-dressing and those which modern knowledge shows to be insupportable can, with absolute safety, be allowed to pass into the realm of mythology. Indeed the main danger which established religions now face is the failure to adapt. They must let those parts known to be false pass into the realm of mythology, otherwise those religions will ultimately become totally discredited. Suddenly they will collapse unless remedial measures are taken. "Ultimately" could well be the near future.

What matters most at the present time is deletion of the command to increase numbers. Safeguarding of moral codes and the maintenance of socially acceptable behaviour patterns come within the responsibilities of religious faiths. The most valuable advance which could be made at the present time is the shouldering of this responsibility by the churches of the world. Now it should be thought a cardinal sin to propagate beyond the means of support. If having families greater than two is officially frowned upon by church leaders, a very favourable effect would accrue. This is the main hope for stopping the runaway population explosion, which, if not checked, could finally extinguish life on our planet. This could happen within the lifetimes of children of the present generation.

At the same time people who have restricted their family size will need to be given preferential treatment. An argument for justifying large families in the Third World is the high infant mortality rate caused by semi-starvation. The response is to increase the family size to ensure that some survive. This strategy will be of benefit to individual parents if resources are uniformly distributed on a per capita basis. The total effect on the entire population is, however, disastrous. The very response to hardship is to ensure matters are made worse for everybody!

Hence the newly-advocated moral code needs to be strengthened by actions which ensure that having small families is seen to be advantageous. Least controversial is the idea of education to make people aware of the consequences of excessive births. But this is unlikely to be adequate in the short time available. Even so this would help without doing any harm. Education might be integrated with project work; people being taken on in relays to share out the work and resulting benefits. This might, for example, be integrated with the establishment of marine farming. Next those who respond by having only two children need to be convinced that they will receive welfare in old age in preference to those who are able to rely on the services of more numerous offspring. This can be coupled with the acceptance of birth control methods by religious leaders.

Then finally if these methods are seen to be inadequate after not more than about a ten year trial period, more controversial, and what at first seem morally unacceptable measures will need to be contemplated. Aid needs to be given differentially so that the small families are given first priority. Or perhaps more politically acceptable would be the distribution of aid on a per-capita basis in which only the adults are counted.

Both of the last two propositions seem very cruel and run into a minefield of controversy. It can be argued that it is not the children's fault that they have many brothers and sisters, so they should not therefore be made to suffer. This is the "fair shares" argument. Unfortunately, in a desperate situation of the kind we will then face, global survival has to take precedence over the desire to be absolutely fair. If fair shares are made the top priority, it seems very unlikely that a solution to the global problem can emerge. People would in any case respond to the measures taken and restrict their families as required, if they knew this improved their chances of survival. Much unfairness would therefore be only a transient problem. It cannot be eliminated and can be justified by the principle of "least global harm". Whatever steps are taken some harm will inevitably arise. Some strategies, those which share out aid fairly but ignore mathematical realities, are totally unworkable in the long term. Ultimately they will therefore do immense harm by allowing numbers to proliferate unchecked.

What is fair is seen to be different by people looking from different angles. A highly simplified view of the world could be considered by way of example. In this the land surface is equally divided and half the world's people are allocated to each half. Initially both have the same per capita food supply and neither has any surplus capacity. There is no further room for increase of the supply. In one half in some way the population is maintained constant by births held at the replacement level.   In the other the growth rate is allowed to remain at 3%, a value corresponding with the average of the Third World today.   This is equivalent to a 2.4 times increase in thirty years.   At this time the growing half finds its population is beginning to starve.   From their viewpoint the other half should share its resources evenly across the world as a whole on a per capita basis.   The half which has taken the trouble to keep its numbers constant does not take this view.   Which view of fair shares is correct?   The only safe way to view the problem is to invoke the principle of least global harm.   If resources are shared on a per capita basis, then within another twenty years the whole world will be starving and soon afterwards the entire population will collapse from starvation.   The alternative will see one half only self-destruct.   On this basis the fair shares solution is unworkable and cannot be adopted.   As a corollary a global solution needs to stress the importance of restricting immigration from countries trying to relieve the consequences of their own growth.   Each nation will have to solve its own population problem whilst contained within its own boundaries.

The best solution will have to accept some unfairness and harm but will minimise the total harmful effects.   But whatever solution is adopted, the action of the world's churches in coming together and mutually agreeing to change their admonitions from increasing to decreasing family size, could be nothing but beneficial. This change would at least cost virtually nothing to implement.

11.7   SEQUEL

After delivering the lecture concerning quantum gravitation to the Department of Physics at Leeds University, I broke my journey home by calling on someone who wished to meet me. He was Douglas Shaw, a powerful direct voice medium.   He had somehow heard about my work and wanted somebody of scientific bent to witness the phenomena in which he was involved.   Through him I conversed with an ex-student of mine whose name was Arthur Cook. I quote from an article subsequently published in "Two Worlds".

MEDIUMSHIP AND PHYSICS - ARE THEY COMPATIBLE?

An entrancing display of flashing blue lights danced in the air. They were attached to nothing yet looked electric. They had beginnings and endings appearing as fuzzy edges, electrical discharges not starting or ending with the usual electrodes. Disembodied lights of short duration in fact, something the like of which I had never seen before. They appear to have been triggered by the rousing songs of Douglas Shaw's circle as they wound up the power. I felt a prickle at the back of my head which others later identified as a breeze. These were the preludes to the most astonishing communication I have ever witnessed.

In a little while a babyish voice began. The sound seemed to emanate from somewhere near the top of the wall, about ten feet from where the medium's heavy breathing came.   It was a little boy addressing his father sitting near me.   A two way conversation developed. It seemed like the ordinary banter between father and son. The medium was not involved in any direct way. The chatter went on for some time, but then a very strong clear adult male voice came through.

The voice addressed me personally as "Ronald" and engaged me in conversation. We talked about the theory of gravitation I had been working on for the last six years and the spin-off which offered explanations for the paranormal. Ultimately he told me his name was "Cook". He said he had been one of the students I had lectured at university. This was soon after my appointment. Unfortunately I was so amazed at the clarity of speech and the implications behind it all, that I just somehow could not seem to think of many sensible questions to ask.

We entered into a technical discussion which seemed just as real as if he were in the same room. He seemed to know a lot about the theory described in this book.

He said that the concept was very close to the truth.
There was an interpenetrating Grid, but it was not
universal as I had assumed.   He said each galaxy was
based on a separate Grid structure.   Also I had the
number of interpenetrating matter systems wrong. At the
time my estimate for the maximum number possible was
ten, though now this figure needs to be revised upwards.
The correct number was seven he said, not ten. He urged
me to press ahead with my technical book called "QUANTUM
GRAVITATION".   I will not repeat what else he said but
his comments were most encouraging.

Another less refined voice started to come through,
addressing everybody in the circle one by one.   He
chided me for not responding and asked if I was asleep.
I wasn't but I was in a somewhat flabbergasted state.
Altogether the evening seemed to me a dramatic
confirmation that we do have other bodies, that our
minds do not reside in the Earth body, and that the
spin-off theory seems to be well supported by the
evidence which can come through the route of mediumship.

I am convinced there was no trickery.   There would
in any case have been no financial incentive involved.
People who still disbelieve this kind of communication
to be possible should read "Keeping an Open Mind" at the
end of Chapter 1.   A healthy scepticism is natural,
nobody can be expected to fully accept these things
without proof.   But if something cannot be disproved,
then at least a belief rating needs to be accredited
which is above zero. To totally disbelieve is to accept
one's mind is fully blocked. Atheists with whom I have
discussed this matter have countered by saying that I am
not justified in rating my acceptance 100% so they are
justified in allocating a zero. To this I reply that I
have been unable to check that Arthur Cook ever existed.
Records at university did not go back far enough and I
do not remember him.   Therefore I cannot allocate the
value of 100%, though I would put the rating above 90%.

This opens a crack in the defensive armour and
allows the thin end of the wedge of doubt to enter,
because it pulls the rug from under the case for

allocating a zero rating. Then time can allow the subconscious to work upon itself with the wedge driving ever deeper of its own accord.

Arthur was very interested in the Grid idea and offered some constructive criticism. In addition he suggested I read three books by an author whose name I have decided not to quote. We will just call him Dr.X. Much of his material was highly relevant and supportive but although I do not doubt the accuracy of other material, this seemed to be too provocative. Part of the objective of this book is to show how bridges can be built across the divide separating science and religion. Much of his material could only increase the barriers.

I obtained his trilogy and read the books with avid interest. Dr.X did not mention any Grid but gave detailed evidence for survival of the consciousness after death. I found the first the most important; the other two are in my opinion too repetitive and add only extra detail in support of the main theme.

It turned out that Dr.X, by observation and without theorising, had offered convincing proof of the existence of a parallel universe to which the essence of all people transferred upon the deaths of their physical bodies. This universe has at least one planet and this interpenetrates our own. Not only that but the personalities of people living in it have changed little as a result of their transfer.

The approach described in the present book, however, started by trying to find an acceptable theory for quantum gravitation. Then wave-particle duality had to be included as this became involved in the solution. Then the only solution compatible with the new theory of gravitation which at the same time was free from unacceptable assumptions, predictions and incompatibilities led to the conclusions already described. Hence, starting from entirely different points and using entirely different methodologies, Dr.X and I home in upon very similar descriptions of reality.

Both approaches, the observational and the theoretical, therefore closely support one another.

This is a very desirable situation and seems to justify the efforts made.

Communication of the ideas described in this book has unfortunately presented far greater difficulty than solving the technical and mathematical problems involved. A final chapter is therefore devoted to analysing the acceptance barriers. Areas needing further study will also be outlined. Much remains to be accomplished in order to achieve a truly "Grand United Theory of Everything". This is the "Holy Grail" of physics at the present time. To achieve this it is necessary to start afresh from a basis having no internal contradictions. The start which I have tried to communicate is free from these and so should provide this base. The achievements of a physics riddled with contradiction simply means that a parallel solution has to exist.

It is my hope that some young physicists will be fired with the enthusiasm to carry on where this book leaves off; to find the answers to the remaining questions.

TABLE T 11.
COMPARISON OF ACHIEVEMENTS - EINSTEIN'S G.R. VS. NEW Q.G.

FEATURE	EINSTEIN-QUANTUM	EXTENDED NEWTONIAN.
1) SPACE	Positive energy states only allowed.	Balance of positive & negative so that net energy is zero.
2) VACUUM PRESSURE.	VERY HIGH NEGATIVE	ZERO.
3) ATTRACTIVE FORCE MECHANISM	Negative coupling. (Contains internal contradiction)	Negative momentum carriers made of negative energy.
4) COSMOLOGICAL CONSTANT.	10^{50} times maximum possible (due to 2)	Zero - As observed.
5) CAUSATIVE AGENT OF GRAVITATION	NONE Gravitons do not fit general rel.ty.	Mediators as 3). Balances mass of attracting object.
6) EXPLANATION of GRAVITY	Curvature of space-time due to nearby massive object.	Compression of space by 5) produces space pressure gradients.

EXPERIMENTAL CHECKS SATISFIED.

7) EQUIVALENCE (Of accln. & Gravitation)	Satisfied exactly.	Satisfied exactly. Force proportional. to total energies.
8) RED SHIFT.	Satisfied exactly	Satisfied exactly.
9) LIGHT DEFLN.	Satisfied exactly	Satisfied exactly.
10) SHAPIRO TIME DELAY	Satisfied by GR.	Predicted to be 1% different from GR.
11) PRECESSION OF MERCURY	43"/century from curved space-time.	43"/century assuming m.v.r conserved.

(For GR this is rigid but in the extended Newtonian physics the true answer will be slightly less since m.v.r is not conserved exactly. This means the new theory is more promising, since with oblateness of the Sun allowed the true value should be 42".)

12) PRECESSION ALL OTHER PLANETS IN SOLAR SYSTEM. Exactly the same values predicted by GR and extended Newtonian.

13) GRAVITY WAVES	Speed = c Energy loss checked	Speed = 0.45.c (Provisional).

(The extended Newtonian physics predicts gravity waves and in the ecliptic of a binary at far distance causes a transverse cyclic oscillation at double the period of the stars. Energy radiation has not yet been formulated)

14) FORCE RATIO	Not predictable.	Fair prediction.

12 A RALLYING CRY!

Non-technical readers will wish to see confirmation of the theory expounded in this book. Not having mathematical expertise they will naturally look to the accepted experts in the field of gravitation for an opinion. They will naturally think of sending the TECHNICAL SECTION to a physicist specialising in this field. Before this step is taken it is necessary to consider some psychological aspects which are only too relevant. Neglect of these will almost certainly lead to a false impression being returned.

The established experts in gravitation are really experts only in relativity and established quantum theory and have not managed to relate the two despite all efforts. They have been taught to accept relativity at a fairly early age and despite the fact that internal contradictions are known to exist experimental verification has reinforced their view that Einstein was right, as shown by Will(124)&(108). They cannot therefore be expected to take kindly to any alternative approach which would undermine confidence in Einstein's theory. To reinforce the barriers, Einstein seems to have become the God of physics, with relativity as its bible. Any attempt to provide an alternative view is regarded as heresy. The "MBM" rejection syndrome described at the end of Chapter 1 will almost certainly come into play and the opinion returned will almost certainly be highly negative. Indeed it is most probable that ridicule will be used to reinforce this opinion. If the request is made that any flaws be highlighted, the most probable answer will be that these are too numerous to mention. This does not mean, however, that there is necessarily anything wrong with the new theory. It is more likely to mean that it has been interpreted as a threat. The greater the strength of the negative response, the greater the threat syndrome is likely to be operating. Therefore in some strange way this helps to indicate that the new is valid. The new is a threat because it is not based on relativity in any way. The experts in gravitation are likely to feel hostile when the validity of a main plank of their specialisation is questioned. If the new is accepted, then they are no longer experts in their field. Understandably this generates a hostile response. One can only sympathise with people in this position and expect to be obstructed by them. It is a situation which has delayed the progress of science throughout history. But sympathy must not be allowed to override the need for rationally-based assessment.

The way to obtain a satisfactory answer is to avoid requesting an opinion. A request should instead be made of a statement of validity of the initial assumptions and for a list of any logical errors and internal contradictions which are found. The next hazard is that the new will be claimed flawed by logic which turns out to be totally false and is readily refuted. What has so far happened with depressing regularity is that untested predictions of Einstein's theory have been used as proof that the new is flawed. Then when it was pointed out that these differences were the very ones which could be used to design experiments capable of discriminating between the two, no further response was ever received.

It can be concluded that the last thing these establishment critics want is to promote experiments which might show Einstein to have been wrong. It is accepted by physicists that new experimental checks for relativity are very hard to find. The new theory, however, points to quite a number. One, the pure creation experiment, has already been described, but the rest are described in the TECHNICAL SUPPLEMENT. Most could be used to discriminate between the two theories. So even if the new is ultimately found wanting, its

communication could achieve a useful role in providing
a new set of experimental checks.

It is important, therefore, that the would-be
refutations are checked for their content against the
list of new proposals for experimental checks given at
the end of Chapter T.S.2.

One most disappointing aspect of the criticisms
received so far has been the lack of comment regarding
successes of the new theory. In no single case have any
of these even been mentioned by any referee. Several
letters have stated that it is insufficient to show that
a single successful check can be met, making reference
to the precession of Mercury. Yet each article
submitted covered all five checks, even stating this in
the summary! Not a single critic has ever written a
word about the way the ratio of the gravitational to the
electric force is predicted. Yet this has not been
achieved in any other theory so far published. It was
a goal Einstein pursued for many years and finally had
to abandon. Surely the open-minded critic should find
this worth just a mention? No single critic has
expressed any surprise that the equations for the so-
called "relativistic" component of precession of planets
is given just as accurately by the new theory as by
Einstein's. Yet there should have been some surprise,
because the equations compared in FIG. 13 (T.S.) look
totally different. Nor has any surprise or credit been
given for the coincidence in predictions given for the
remaining checks. There has been a total absence of
comment regarding the internal inconsistencies shown in
Chapter T.S.2 to be inherent in special relativity.
Hence none of the criticisms received to date can be
regarded as unbiased.

To obtain an unbiased judgement it is necessary to
go to people who are not too closely involved. Nobody
needs to be an acknowledged expert in gravitation to
give a value judgement. Indeed experts need to be
avoided. Anybody familiar with Newtonian mechanics and

having a background in "A" level mathematics will be able to judge the issue. Suitable people are professional mechanical, electrical and structural engineers and mathematicians outside the field of cosmology together with young physicists. Young physicists have not yet invested so much intellectual capital in the established approach as to make them feel reluctant to explore alternatives.

This advice is given as the result of a frustrating six years of effort in attempts to publish aspects of the new theory in the scientific journals. Not a single success has so far been achieved, yet none of the assessors has pointed out a single valid objection. I will quote extracts from two examples to show the kind of objections which have been received:-

25 Aug. 1987

"In selecting papers for consideration we have to apply our editorial judgement as to whether the arguments contained in the paper concerned will have an immediate impact on the readership of Nature - on the course of research in particular. This is not to say they should not be published at all. It is simply an editorial decision that they cannot be published in Nature."

Several articles had been submitted over about a three-year period regarding gravitational topics. These included Tryon's "negative gravitational potential energy", the need for negative mass as mediators for attractive forces and a solution to the long-standing problem of a huge "cosmological constant" predicted by established theory and known to be totally false. Several redrafts had been made in each case to satisfy previous objections, which were all of a trivial kind.

Yet reference to the last problem is made several times in the workshop proceedings of 1982, edited by Gibbons, Hawking and Siklos(207). They repeatedly and openly state that failure to resolve this problem undermines confidence in their theory. They are trying to so amalgamate general relativity with quantum theory that the universally attracive force of gravitation is predicted. Instead the combination results in the prediction of a huge **repulsive** force associated with a "cosmological constant", which is said to be 10^{50} times greater than astronomical observations can possibly allow! In Chapter TS1 the mathematical logic involved is summarised being taken from the book by Novikov(211). It attempts to amalgamate three theories each based on assumptions which are incompatible with the others. This goes against all the rules of logic.

Then a year after my attempts to communicate a solution an article appeared by Abbott(101) in the same journal, addressing the same problem. Briefly he says that there are other universes existing in higher dimensions and are connected to ours via "worm-holes" in space. It is then hoped that they have a similar problem of a huge cosmological constant which somehow is exactly equal but opposite to ours so that mutual cancellation takes place. The article ends by stating that "the approach relies on a shaky formalism and on many untested assumptions."

At present quantum theory has no satisfactory explanation for attraction. It relies on the idea of "negative coupling", which, as shown in Chapter 5 of this book, is an untenable concept because it contains another internal contradiction. The rejected article showed the necessity for accepting the existence of negative energy states and then went on to show how space, regarded as an exact balance of positive and negative states, would predict the cosmological constant to be zero as required. Again this is summarised in Chapter TS1. Most of the assumptions used in new concept have been fully tested and accepted in mainstream physics. They are listed in Chapter T.S.1 so the reader will be able to judge the issue.

Why do journals accept solutions which openly state their logic to be highly questionable with initial assumptions untested or untestable, which also involve internal contradictions, and yet reject solutions which fit the facts without involving contradictions?

It is not my wish to discredit "Nature". Indeed this particular journal has been one of the most reasonable in the way it has dealt with my submissions. I have received from them references for books I should consult on the subject and they have always taken great pains to find reasons for refusals to publish. They cannot be blamed for this reaction. Clearly they have to retain their readership. They know what their readers will be pleased to see. Clearly a theory which would "rock the boat" is unlikely to please.

This puts the finger on the crux of the matter. It shows why new ideas are always resisted in the initial stages to the detriment of scientific advance. The barriers in the way of progress seem virtually insurmountable.

Some journals refused these and other articles on grounds that they were not specialists in gravitation, despite the fact that they published articles relevant to this topic from time to time. They advised sending the work to specialised journals. The following is typical of the specialist response:-

CANADIAN JOURNAL OF PHYSICS

"QUANTUM GRAVITATION" (Submitted 8/8/88)

"This paper does not meet the standards of CJP, since it fails to properly connect with currently accepted gravitational theories such as Einstein's general relativity and the presently accepted problems in quantum gravity."

The new theory completely resolved the problems mentioned because they simply did not appear in a theory quantum-based from the start. The paper showed that the extended Newtonian physics gave all the same predictions as Einstein's theory except for gravitational waves, which had not been studied. This final rejection had been preceded by several acknowledgement cards saying that the paper was under review during the 16-month period they took to make up their minds. Clearly they were unable to find a logical flaw to justify rejection.

Nobody has yet been able to connect Einstein's general relativity to quantum theory. Hawking(115) openly states on page 11 of his best-selling popularisation that quantum theory and relativity are now known to be incompatible with one another, so one of them must be wrong. He implies later that relativity is the theory chosen as correct. So if Einstein is wrong, and the present work suggests that he is, there is absolutely no official way of communicating alternative theories. Clearly the rejections are all motivated by political reasons which are being allowed to totally override scientific evaluations. The assessors decide in advance what is acceptable on the basis of its compatibility with established views; then if it does not fit they look for excuses no matter how weak, or use false logic to justify rejection. This is just very bad for science. Science can only thrive in an atmosphere of constructive criticism. All new ideas which have been checked and found free from logical flaws need to be positively encouraged, otherwise progress is stifled.

My next approach was to offer the new theory to a number of universities, in turn, as a Ph.D. thesis. Already having an honours degree in a scientific and mathematically based discipline, university regulations are already satisfied. I thought if I had a Ph.D in physics, then the acceptance barriers might no longer be insurmountable. It is tempting to name the universities which have been approached but this will be resisted. There is no need to go out of one's way to cultivate enemies! One refused to look at it. Another used the excuse that they would have no time to assess it owing to pressure of other work. So they also would not look. A third said that their rules forbade anyone submitting who had not obtained a first degree at that university. I do know they can waive this rule when it suits them and they are also aware of the new approach.

At least I know what Galileo must have felt like when nobody would look through his telescope!

When excessive time is being taken to solve an intractable problem it makes sense to diversify effort

by studying all possible approaches. Other disciplines
can help by offering new angles. It could be that in
the area of gravitation and cosmology theoreticians have
been following a false trail for the last seventy years.
Hawking(115) states that he thinks they will have a
solution for quantum gravitation by the turn of the
century. So they still think they need ten more years!
When this time is up, will they still need ten more
years? Will they ever reach their goal following their
chosen path? Could it be that with every move they make
they depart further from reality? These are the
questions which in my opinion the critics should be
asking. The theory Abbot described (which was
originated by S. Coleman) and which is summarised above
indicates the way the thinking is developing and the
reader can judge its plausibility.

Science needs something like the Hippocratic oath of
the medical profession if it is to be prevented from
stagnating. This should state that all new ideas, from
whatever source, be given a fair analysis. Ideas need
to be criticised to discover any fundamental flaw and if
none can be found, then without exception they should be
encouraged and receive priority for publication,
regardless of any opposition from others who might
consider their interests impaired.

If a group of professional people like me with
expertise in mathematics, Newtonian mechanics,
electromagnetics, fluid mechanics and thermodynamics got
together with some young physicists who know more about
the strong and weak forces than I, then we would stand
a good chance of beating the acknowledged experts to the
"GRAND UNIFIED THEORY"! This is the "Holy Grail" of
physics at the present time. Some more work is needed
to find a way of calculating the energy loss involved in
the radiation of gravitational waves. Then some
modification may be needed to quantum theory to express
the strong and weak forces in terms of three spatial
dimensions plus time and energy. We need to explore
established quantum theory to find all the areas in
which negative energy states are causing embarrassment.

As shown in Chapter 5 negative energy forms have to exist.   Hence what is indicated is some wrong assumptions or false theorising. We will need to sort these out and come up with viable alternatives.

The major point is that whatever successes are claimed for the established approach, it can never be valid because it contains so many internal contradictions.   Even a single contradiction is sufficient to invalidate any theory! These successes therefore have to be interpreted as meaning that a parallel alternative solution must exist and needs to be sought.

12.1 MEETING CRITICISM

A constructive criticism was recieved from Richard Austin and is of the kind I like to see. He is a competent mathematician familiar with quantum and relativity physics. He stated that the great attraction of the route being followed by mainstream physicists is that there are promising signs that they can eliminate the need to put universal constants for the forces of nature in "by hand". It would not be possible to do this with the approach made here. This may true up to a point, but an experimental determination of at least one constant has to be made somewhere. The objection is at least partially refutable by the theories given in the TECHNICAL SUPPLEMENT where it is shown that from a common starting point the same energy densities of space are required to relate the electric and gravitational forces. The lower densities required by the strong and weak forces can readily be accommodated. The new approach might be capable of extension to achieve the same promised result. This aspect has simply not been explored. The main emphasis has been to find a common base free from the contradictions bedevilling the established approach. Until a theory is soundly based refinement seems pointless.

He also thought my splitting of the four forces into two groups would be cause for criticism. I place the electromagnetic and gravitational forces into one group based on buoyancy type forces. These are caused by the virtual particles of space forming energy density gradients. The strong and weak forces fill the other group, since they operate by mediator absorption processes. The reason for this is fully explained in the text. In trying to use a common absorption basis as in established theory it seems that some important factors have not been recognised. It is not possible to explain the absence of gravitational shielding by layers of matter, if mediator absorption models are postulated. Furthermore such models will predict both planets and electrons to progressively gain angular momentum and spiral out of orbit. For the strong force, speeds approach that of light and then the objection does not arise. In this case, however, absorption is necessary since binding energies would not otherwise be reflected as mass deficits.

He pointed out that a major triumph of modern physics was the successful relating of the electromagnetic force to the weak nuclear force. This suggests that the same absorption model will apply to both. By sheer coincidence, however, it turns out that for the electric force almost the same energy density of space is needed for the absorption model as is needed for the buoyancy type. A parallel solution is therefore indicated which is free from contradiction with observation. A common absorption model does not need to obtain.

It is much more important that a theory is free from internal contradiction and fits observation than that it fits in with what theorists feel to be a neater arrangement. None of the accepted theories meet this criterion. However, in the present theory all four forces depend on a common quantum base so there is still a unification of the four forces at a basic level.

He also felt that having a Grid which deliberately designed the universe to permit life forms to exist would lay the theory open to criticism. After all, he says, Jews and Christians say the world is as it is because God made it that way. True. But find a way of explaining wave-particle duality which does not have inbuilt paradoxes, uses unacceptable assumptions, violates causality and makes true prediction without a universal computing base interpenetrating the whole of space. Find one which explains why atoms are ball shaped and not like discs without being based on a Grid. If these matters can be satisfied without one, and no theory has emerged which can, then and only then, does such a criticism hold water. With a Grid allowed then a creation scenario seems perfectly acceptable. If this is how it all really happened and scientists are totally committed to theories based on a purely accidental origin, then it is impossible for a solution ever to emerge.

The point of this presentation is to show that an alternative is plausible and should be pursued in parallel with the accepted scientific route. There is no way any religious group can be proved wrong in the common fundamental belief in a Creator and so this needs to be considered as a viable option which needs to be explored. It must also be repeated that the present solution did not start out to prove God exists. I was an atheist when I began. This was homed in on as an inescapable conclusion.

It is also true that no attempt is made in the present theory to predict the existence of new particles such as the "W" and "Z" which were a triumph for the existing approach. My answer here is that this part of quantum physics needs to be integrated with the present theory instead of trying to fit it into the far more abstract and totally incompatible curved space theories of higher dimensions. As Mr Austin points out in the supportive part of his critique, the ideas of space curvature and mediating particles are basically incompatible. Either mediators are needed to produce forces or particles can be assumed to feel accelerations caused by curved space. These theories cannot be satisfactorily integrated because each provides a complete explanation without need for the other, he states.

So another avenue is opened for extension of the new
theoretical approach which started by extending
Newtonian physics.

Mr Austin thought the material would prove very
heavy reading for the layman. He did not consider the
presentation would have a popular appeal comparable with
a Paul Davies book. This I have to ruefully admit is
probably only too true. I too have found the Paul
Davies style captivating. I have done my best but I am
not a naturally gifted writer for the non-specialist.
I made great efforts to follow the accepted route of
first obtaining scientific acceptance, only to find all
doors locked and barred. I have therefore to try the
popularisation first. It is the wrong way round but I
am left with Hobson's choice.

But his criticisms also had a strong positive
aspect. He said he was fascinated to see a theory
capable of explaining paranormal phenomena. This was
something he had hoped to see for many years.

This is how a critique should appear. It represents
constructive criticism. The kind which avoids
mentioning positive aspects is purely destructive and
needs to be condemned.

There is also another area in which more study is
indicated. This will now be considered.

The abstract mathematical reasoning which led to the
formulation of electromagnetic wave theory also needs
inspection. In Chapter T.S.1 a physical structure is
briefly described which relates the photon to space so
that this wave can be imagined. This approach indicates
that in the formulation, a phase lag of $\pi/2$ must have
been dropped somewhere. In the philosophy on which the
present theory is based all mathematical formulations
need to be consistent with imaginable logic. Any
mismatch which arose during development of the extended
physics presented in this book was taken to indicate the
existence of an error in either the initial assumptions
or the mathematics. A flaw was always found and

rectified. The place where the $\pi/2$ has been dropped will ultimately be discovered.

This highlights the difference in the present approach from that of contemporary physics. This is best expressed by mentioning another of the many rejection letters received. My submission said that the new theory had the advantage of being easily imagined. Instead, the assessor used this as his reason for rejection. He claimed that physics had become so sophisticated that it had long passed the stage at which anything could really be imagined. Physics had developed into pure abstraction. Therefore this was indeed a valid reason for rejection because it indicated the submission to be a regressive step. A viable solution to the problem of the "Cosmological Constant" was unacceptable because it did not look difficult!

Surely simple approaches ought to be communicated so that people can criticise them and compare with the established view. Theories involving contradictions or so-called paradoxes means that science is going up a blind alley. Are students being asked to take on board theories containing internal contradictions which logic should prohibit? In this book, in both Part I and Part II, internal contradictions are shown to exist in the established approach. If students are being trained to accept false logic then a situation exists which needs to be rectified. Each new generation will continue to mislead the next unless a group from outside the discipline can develop sufficient credibility to override inbuilt inertia.

If a new group can complete the picture, then it would indeed achieve the "Super Grand Unification" which combines within it the explanation for psychic phenomena. The Cambridge Professor, Brian Josephson(116), argues that this has to be the next step. It would not be as difficult as it might seem at first sight. The new theory already relates the magnitude of the force of gravity to the electric force and this has not been done before. Already a theory is offered here which integrates gravitation with an

explanation of psychic phenomena based on the evidence
of wave-particle duality. Also, as shown in Chapter 3,
the idea of higher spatial dimensions arose because
theoreticians could not explain gravity by accepted
Newtonian methods. The latter used only three- length,
breadth and height. But the extended Newtonian physics
succeeded without using any more.   Hence the
justification for higher dimensions no longer exists.

The young physicists we need may be able to reduce
present explanations for the strong and weak forces to
the same terms. Present-day theoreticians in this field
could be experts in false concepts, because according to
Hawking(115) they are developing the new "string
theories" in either 10 or 26 dimensions. Consequently
it is highly probable that their formulations do not
represent reality. If so the door is wide open.   It
would also be necessary to find an amateur journal in
which the work could be published, because it is quite
clear that none of the professional journals would be
interested.

The information would be trapped for some time
within a small group, since the establishment would not
accept the logic. The chances are that a completely
successful theory of everything will emerge based on the
extended Newtonian approach. It would spread out from
the centre and could ultimately become recognised.

Just one warning note needs to be sounded which the
young physicist would be well advised to heed. In the
past people coming up with new ideas which contest the
established view have injured their career prospects.
This happened in the case of the emerging theory of
"quarks".   For protection, therefore, the young
physicists who think as I do and want to join in, had
better not disclose their names. Instead, a system of
numbers can be allocated. Then when sufficient strength
has been acquired, and not until then, names associated
with those numbers can be disclosed.

13   CONCLUSION

The picture of a possible multi-universe system based upon an all-pervading Grid network having computer-like properties has been painted. For the most part it acts like a machine, being totally amoral. However, it has more highly developed centres which allow emotion to be experienced and these are also the places where a will and a creative urge reside. The entity is the Creator of the illusory universe which we observe and of which we are a part. This Creator is not a bearded old man in the sky, however, but a life form permeating all matter and space. The bearded-old-man image is merely a useful construct of the human mind invented to help us relate to the "Infinite". We are not built in the physical image of the Creator because this has no need of the limbs which are appropriate to Earth. However, and more importantly, we reflect the same emotional needs and creative urge. We are indeed extensions of the Creator, having an important function and purpose.

Because atoms in reality occupy a huge amount of space in relation to the volume of all their constituent sub-atomic particles put together, it is possible for a number of matter-systems to be supported by a common Grid. They can interpenetrate one another without resistance because their sub-atomic particles do not couple with those of any other matter-system. Life-forms can therefore be composites living simultaneously in several interpenetrating systems of matter. The complete picture then shows that extended physics is consistent with basic religious beliefs. These are the existence of a Creator and of the soul or consciousness. The latter is a separate entity which interpenetrates the Earth-body whilst it lasts, using the brain as an interfacing component. The whole supports the belief that the consciousness survives physical death and indicates that life has an ultimate purpose.

There could be more than ten interpenetrating universes operating in the same manner, the limit being the point at which the virtual particles of space become jammed into a solid mass. Each system of matter operates independently of the others except for shared gravitation. The other universes probably depend on our gravitation for holding the matter of their planets together, because their reported properties are consistent with matter-systems constructed from exactly balanced positive and negative physical energies. Matter of this kind would exhibit zero inertia and so could be accelerated almost to the speed of light with very little effort. Such systems cannot, however, produce any gravitational force by themselves.

Normal and paranormal phenomena were shown to have a common basis. They both depended on two kinds of physical energy, positive and negative, which formed the building-blocks of systems of matter which could interpenetrate one another. The computing power of the Grid was finally pinpointed as defining the nature of "psychic energy". This is therefore an abstract form and so cannot be measured like the positive and negative physical forms of energy. Psychic energy, force, power, call it what you will, is a power of information which uses wave-functions as numbers. These are manipulated by the underlying Grid for the organisation of physical energies into the structure matter.

Paranormal effects like telepathy, psychometry, dowsing, spoon-bending, the apport, spiritual healing and the like then became explicable in terms of the computing power of the Grid. Evidence of psychic effects, which have been observed, show that the human mind has the capacity to penetrate right into the hearts of atoms. There it senses the tiniest tip of a huge iceberg of information stored in the memory banks of the Grid.

THE MIND IS SEPARATE FROM THE BRAIN!

Psychometry works! It is a fact as far as I am concerned since I have proved it experimentally to myself. Others who doubt its reality should test it for themselves. And this provides dramatic evidence that the mind is a separate entity. To reach down into the Grid whose filaments cannot have a spacing more than 5% of the radius of an atom, a mechanism of similar grain size needs to be involved. The neurones of the brain are many thousands of times too large, each being composed of many thousands of atoms. Hence the brain by itself could not possibly explain the workings of the mind in terms of neuronal networks as most parapsychologists believe.

Even the other interpenetrating "spirit" bodies of possibly finer grain structure would face the same objection. Their matter systems have to be based on the same Grid and therefore must also be of larger grain size than the Grid itself. Hence only one possiblity remains. The mind, the seat of consciousness, needs to be a part of the switching pattern of the Grid itself. In the ultimate state, shed of all its clothing of matter-systems, it could exist as an independent life-form!

THE NEW PHYSICAL BASIS

This picture emerged from a study which aimed to explain why sub-atomic particles appear to travel along all possible alternate routes simultaneously as if they were spread out like waves. The solution showed that waves merely acted as the control system, used like numbers, to specify where sub-atomic particles were to be re-created. Such particles, which established physics assumes to be permanent, turned out to have transient lives, just like the virtual particles of space. They are instead particle sequences joined end to end in time but not in position. They are based on a permanent bit of physical energy which keeps on being absorbed by the Grid to be used for re-construction and in between times can travel at speeds many thousands of times the speed of light along its filaments.

In this way an explanation emerged for the way the electron can dodge about all over the ball or lobe shaped orbital of its atom. This is the shape specified by Schrödinger's wave equation. It shows why the atom of hydrogen can appear ball-shaped. It would appear as a flat disc if the electron were a truly permanent structure.

The explanation is consistent with the new theory of quantum gravitation which satisfies the experimental checks at least as well as Einstein's Theory of general relativity, being simultaneously free from its difficulties. In the new approach it is the compressibility of space producing buoyancy effects on sub-atomic particles which has the same effect as Einstein's curved space-time in providing equations which correctly predict observation.

Only gravity waves are excepted because this part of the study has not yet advanced very far. The theory does, however, predict gravity waves, but the energy they carry away has not yet been formulated. A much more important plus for the new theory is that it predicts the ratio of the magnitudes of the electric to the gravitational force to within striking distance. It is the first time this ratio has been predicted. Einstein spent many years trying to do this from a base of general relativity and totally failed.

The new theory is compatible with the idea of negative energy states. Negative states were proved to exist by showing that forces of attraction could not otherwise be explained in a satisfactory way.

SPIN-OFF

Exciting spin-off from the new solution to the problem of quantum gravitation appeared. It might provide an important tool to help in our fight to save the planet. This is because, just as Einstein's theory threw up the possibility of a new energy form as power from the nucleus, so the extended Newtonian physics predicts a new energy form. It shows that gravitational potential energy is not a true energy form at all. It is a pseudo-energy form. When an object is rising by its own inertia its speed and associated kinetic energy are continually reducing. There is a continuous destruction of total energy in consequence. This is exactly balanced, however, by the destruction of an equal amount of the negative energy of space. Energy is not "stored" in space as textbooks say. On falling again both energy forms are re-created in exactly balanced amounts.

Hence the new theory indicates that the pure creation of energy is feasible, despite its prohibition by the so-called "First Law of Thermodynamics". This law is only a special case arising when balancing negative energies are held constant. The new concept allows the universe to have been created out of nothing in the first place. It also indicates that it should be possible to find a way of generating energy tomorrow by pure creation. An experiment was proposed which ought to confirm feasibility. At the same time a positive result would support the new theory of quantum gravitation and show general relativity to be falsely based. It would lead to the abandonment of theories dependent on the idea that curved space-time and higher spatial dimensions can exist.

Energy by pure creation would be available at any point, there would be no need for carrying fuel and, best of all, there would be absolutely no pollution of any kind. It might be the answer to the global greenhouse problem. As a bonus it would be impossible to create explosions or other harmful effects by pure energy creation. Hence a wholly benign solution to the energy problem may emerge.

NEW EXPERIMENTS AND MODIFIED RELATIVITY

Another six new experimental checks were proposed, most of which would discriminate between the new theory and Einstein's "special relativity". In Einstein's theory, light propagates independently of space when travelling in a vacuum and this assumption is incompatible with the base of any conceivable quantum

theory.   It is particularly incompatible with the
present quantum-based theory of gravitation.   In this,
as described in Chapter T.S.1, light is a complex
dynamic structure of space and is therefore totally
committed to propagating *relative to space.*   Space being
a fluid, however, has different speeds at different
places.

The Earth, according to the new theory, carries
along a bubble of space having a radius reaching nearly
to the moon.   The Sun carries along a vastly larger
bubble enclosing the solar system and Sun-space flows
around the Earth bubble as it moves in orbit.   Even
galaxies carry bubbles, so bubbles exist within larger
ones until the whole universe is enclosed.   Any other
motion then needs to be measured relative to the local
space enclosed by a bubble whose edge defines a sudden
change in the speed at which space is moving.   The
choice of reference frame is no longer arbitrary as in
special relativity.   The TECHNICAL SUPPLEMENT shows how
known experimental checks can now be satisfied with
light propagating relative to local space.

PHYSICS GIVES SUPPORT TO BASIC FAITHS

As already stated, the combined solution to the
problem of wave-particle duality and gravitation threw
up support for the basic tenets of the religions of the
world: the faith in a controlling intelligence or God
and the belief in a spiritual existence.   To the
religious I would hope this will imprint a message.   It
could help draw the religions of the world together, to
shelter under the umbrella of extended physics.   There
should no longer be any need to defend belief against
the discreditment of science.   It simply does not appear
possible to explain wave-particle duality and
non-locality by theories which do not involve a Creator
permeating all things.

An answer to the question "How could God create
Himself?" has been suggested by an evolutionary
scenario.   Also "Why were we created?" was answered by
the suggestion that we are extensions of the Creator,

the arms and legs and other adaptations which permit the expression of creative need.

It should be possible for science and religion to blend together and evolve as one. There can be only one truth. So in principle the discoveries of science need to be in harmony with religious faith. Any discord means something must be wrong somewhere. It is my hope that the solutions given here will be interpreted to show that discord can be banished. If this is accepted, then no longer does any justification exist for the stagnation of religious thinking, which has served to defend it against the advance of science over the last three hundred years.

But all religions accumulate much that should be jettisoned. Up to now it seemed that all this had to be retained. Admit some parts faulty and the whole loses credibility. But this applies when faith is the only supporting prop. Once religious faith is accepted to be supportable by physics, there is no longer any danger of discreditment. Then that which should be jettisoned can be jettisoned. This needs no formal admission of fault. All that is required is to simply ignore certain material and allow it to pass naturally into the realms of mythology. Just by concentrating on the parts which matter, progress can be made.

Unnecessary conflict exists between different religious sects, but what is right for one may be wrong for another. For example, the established churches are probably right to discourage dabbling in the occult for the majority of their flock, since this might involve danger. They should not, however, attempt to discredit those who feel a need to communicate with people of our companion planets. Such individuals are probably more advanced and need such activity. They would not be prone to the same dangers. They are also needed to provide the credibility which only observation can supply. This can support both theoretical physics and religion. A neat triangle of disciplines acting in mutual support.

These things are important now. We humans, by

failing to control our numbers, have become as a plague
of locusts, destroying our planet with reckless abandon.
This is where healthy world religions need to bring
their creative force to bear, for this is their area of
responsibility.   At present their doctrines mostly
encourage people to have large families. They need to
revise such admonition. They need to substitute the
idea that it is a sin to procreate to excess. Then in
collaboration with mathematicians and social workers
they need to use their powers to help implement measures
which will effectively reduce the birth-rate to the
replacement level within twenty years.

The planet can be saved only by a global combined
effort. Harmony in science and religion could greatly
help in the winning of the battles to come. This does
not refer to warfare. The fight to save the planet is
of a non-military nature.   But to win through will
demand more dedication and sacrifice than all the wars
in history combined.

NOTE.

*People wishing to have the technical part of this
book checked by experts are advised to read CHAPTER 12
before proceeding.   Unless the psychological factors
discussed there are taken into account, it is unlikely
that an unbiased assessment will be returned.*

PART II

TECHNICAL SUPPLEMENT

(T.S.)

QUANTUM GRAVITATION

by

R. D. Pearson

CHAPTER T.S.1 is an extended version of the lecture given at The University of Leeds on 18 January 1990. It deals with the EXTENDED NEWTONIAN concept for explaining gravitation which starts from a quantum base. The problems existing in established theory do not arise.

CHAPTER T.S.2 deals with a modified concept of special relativity which avoids the internal inconsistencies of Einstein's theory. It also avoids a basic inconsistency with the new quantum based theory of gravitation.

Seven proposals for new experimental checks are included.

CHAPTER T.S.3 deals with the problem of finding a physical model for the electron to explain its properties. From it the energy density of space is deduced and this enables the ratio of the electric to the gravitational force to be predicted, at least approximately.

CHAPTER T.S.1

QUANTUM GRAVITATION

LECTURE AT THE UNIVERSITY OF LEEDS   18 JAN 1990

1.1   INTRODUCTION

My aim is to show that a satisfactory quantum-based theory of gravitation can be formulated simply by extending the Newtonian in quite a simple way. The predictions made exactly parallel those of Einstein's "general relativity" and also meet the experimental constraints of his theory of "special relativity". Yet the derivation is not fundamentally based on relativity in any way. Conversely heavy reliance is made upon quantum theory. This is not unreasonable for a theory which is quantum-based from the start.

After formulating his two theories Einstein spent a large part of the rest of his life trying to relate the magnitude of the electromagnetic force to that of gravitation. He totally failed. In the new approach it will be shown that these are related automatically to within a few orders of magnitude. Then the difference is used to determine an important constant. Hence a very big step toward the much-coveted "Grand Unified Theory" seems to have been made.

I am also going to show that this approach avoids the difficulties which arise when attempts are made to match quantum theory to relativity. One problem arises from inconsistencies caused by the breakdown of logic at the "event horizon" of the much-discussed "Black Hole", which relativity predicts.

A main stumbling-block is the so-called

248

"cosmological constant". This arises from attempts at amalgamation and is associated with a huge and totally false predicted force of repulsion pushing the galaxies apart.

Stephen Hawking(115) says that quantum theory and relativity are known to be inconsistent with each other, so that one of them must be wrong. He also implies that the approach being taken assumes Einstein to be basically correct, so that quantum theory is having to be modified. Forces are being explained by curved geometries requiring higher dimensions for their existence. Hawking states that the new "superstring" theories are being written in no less than 26 dimensions.

But mathematicians have been following this route for several decades without achieving a theory capable of predicting all known observations simultaneously. This, together with the requirement that no internal inconsistencies exist, are the criteria for success. Hawking says that he does not think a solution will be available until the turn of the century.

Is it not time, therefore, to try the other route - assume a quantum theory written in only three spatial dimensions, based on Euclidean geometry, but with Newtonian physics extended a little? As will be shown, time and energy also need incorporation as separate dimensions, so in total the approach can be regarded as five-dimensional. Such an approach means abandoning relativity in its entirety and also all idea of representing forces by curved geometries. Yet all the tested predictions of both relativity and quantum theory need to emerge from the new approach. It seems an exceedingly difficult target to reach, yet a solution will emerge based on common-sense and without recourse to very difficult mathematical concepts.

In effect we return to the first decade of the century, to the point physicists had reached, trying to explain the mysteries of the universe by the now-derided "Clockwork of Newtonian physics". I feel they should have kept this route open, exploring all possibilities

at once like the photon or the electron. It is simply
unscientific to have everybody following the same path,
because initially there is no way of knowing it to be
the one which is correct. Chapter 5 of the non-
technical section provides an extension to this
introduction.

1.2 PRIMARY DERIVATION
1.2.1 ASSUMPTIONS

The assumptions first need to be listed. These
are:-

1)   Only three spatial dimensions to be allowed, written
in the straight-line geometry of Euclid, with time and
energy as separate dimensions. Total number of
dimensions = 5.

2)   TOTAL ENERGY "E" is to be defined as the sum of the
rest energy "E_o" of an object and its kinetic energy
"δE". Any binding energies are automatically included
in the rest energy but **POTENTIAL ENERGY** is **NOT** included
in the term **TOTAL ENERGY**.

3)   The entire theory is to permit of visualisation
without recourse to analogy; it is a common-sense
approach.

4)   Newton's law that force equals rate of change of
momentum is to be assumed to apply exactly, i.e.:-

$$F = \frac{d(m.v)}{dt}$$

5)   The quantum vacuum is to be assumed as a seething
mass of virtual particles. Some are the mediators needed
for transmitting the forces of nature and these move at
the speed of light "c".

6)   The gain in energy measured in a given direction is
equal to the product of the net force and distance moved
in that direction, i.e. "$\delta E_x = F_x.X$" etc., where "δE_x" is
the component of kinetic energy measured in the "X"
direction.

7)   The net energy of our universe plus space is always
zero.

8)   Negative energy states are not disallowed.

9)   A protective barrier exists around particles so that when those made from opposite kinds of energy interact they do not mutually annihilate immediately.   The barrier works both ways so that, in addition, particles are prevented from emitting energy forms of opposite sign to themselves.

10)   Linear and angular momentum cannot be interchanged. Then "spin" cannot affect the impulse produced by mediator/particle interaction, though it can determine the interaction "cross-section".

The last assumption has been introduced because in established theory the "graviton" is the mediator of the gravitational force.   It has positive mass-energy with spin 2.

It is therefore a positive momentum carrier but the spin allows it to produce an attractive force instead of producing the repulsion it would otherwise exert.   A very large proportion of the total energy could be associated with the high angular momentum represented by this spin, the remainder being the kinetic energy of linear motion.   On interaction with a sub-atomic particle it is conceivable that all the rotational energy could be transformed to the linear kind to increase the linear momentum of the graviton.   It could pass through the target and, by speeding up, would then produce an attractive response.   The target particle would be impelled in the direction of the particle which first emitted the graviton.

However, this is the only way the direction of force could be reversed without introducing the internal contradiction associated with the "negative coupling" described in Chapter 5.

The interaction would eliminate the spin and so the emerging particle would no longer be a graviton able to exert any further attractive response. It would be effectively used up in the interaction. As shown later, if mediators are absorbed or used up by producing attraction,   then   predictions   result   which   are   at variance with observation. This is the reason why assumption 10 has been included. It also follows that the graviton cannot be incorporated in the present approach. Attractive forces demand mediators consisting of negative energy so that they carry negative momentum.

1.2.2 THE NEWTONIAN STARTING POINT

It is possible to start entirely from the Newtonian assumptions and derive the important equations which Einstein obtained by his concept of "special relativity". This is left to Chapter T.S.2 however. For the present a start will be made by assuming that Einstein's method is followed exactly for motion at speeds comparable with that of light in zero gravity.

The same conditions will obtain for horizontal motion in a gravitational field, provided every particle considered is acted upon by a hypothetical force, which is exactly equal and opposite that of gravitation. Then motions starting out in the horizontal direction with speed "v" will remain horizontal. The upwardly-directed speed component "u" will be zero at all times and "v" will remain constant. Hence the following equations obtained from the copy of Einstein's method will apply at all levels in the field. They are:-

$$E = E_o \cdot \frac{1}{\sqrt{1 - \left(\frac{v}{c}\right)^2}} \qquad [1.1]$$

and:-

$$E = m.c^2 \qquad [1.2]$$

1.2.3 BENDING OF A LIGHT BEAM

Next let us consider the bending of a beam of light in a gravitational field. This is shown simply in FIG.1 as a circular arc, although a parabolic arc would be more exact. It makes no difference to the outcome. This is proved in the author's general derivation given in the full book(212).

If light is bent in the field, and if it is controlled by waves, then photons must always move in a direction perpendicular to the wave front. Hence the

speed on the outside of
the bend must exceed
that on the inside so
that waves keep in
step. This assumes the
observations are made
from a fixed datum.
This is important
because, as will be
shown, the units change
with gravitational
potential.

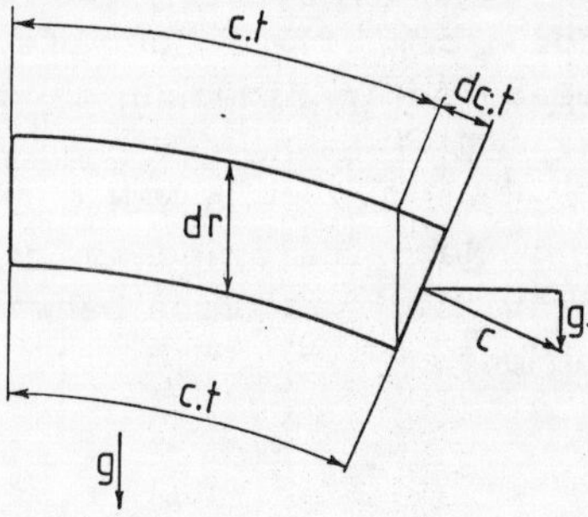

$$\frac{g.t}{c} = \frac{dc.t}{dr}$$

$$c.dc = g.dr$$

From the velocity
triangle, starting out
horizontal, a vertical
speed "g.t" will
develop in time "t",
when the universal
acceleration "g" is
applied with light
behaving like other
matter.

FIG. 1 GRAVITATIONAL

LIGHT BENDING

In the same time
the outside of the
curve must exceed the distance on the inside by amount
"dc.t" when a distance "dr" separates the inside from
the outside. From similar triangles it follows that:-

$$c.dc = g.dr \qquad [1.3]$$

The idea of using a circular arc in this way was
taken from a popularisation of Einstein's theory given
by L.C.Epstein(111). He used the alternative deduction
of assuming the speed of light to remain constant so
that time had to be the variable instead. In this case
the datum position had to be located at the place where
"c" is defined. It had to move with the level of the
object to which reference was made. Either

interpretation is equally valid but for present purposes it is necessary to fix the datum at some position "1" for "c_1", then "c" becomes a variable and time is measured in fixed units defined at position "1".

1.2.4 A NEW QUANTUM GRAVITATIONAL LAW

Next it is necessary to modify Newton's law for the gravitational force between two spherical well-separated objects.

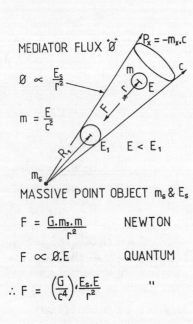

MEDIATOR FLUX "ϕ"

$P_x = -m_x.c$

$\phi \propto \dfrac{E_s}{r^2}$

$m = \dfrac{E}{c^2}$

$E < E_1$

MASSIVE POINT OBJECT m_s & E_s

$$F = \frac{G.m_s.m}{r^2} \qquad \text{NEWTON}$$

$$F \propto \phi.E \qquad \text{QUANTUM}$$

$$\therefore F = \left(\frac{G}{c^4}\right).\frac{E_s.E}{r^2} \qquad "$$

FIG.2 GRAVITATIONAL FORCE

The new quantum-based theory is illustrated in FIG.2, showing a cone of mediators emanating from a massive point object. Since mediators must spread out over an area increasing as the square of radius, then an inverse square law will apply to the mediator flux density. Flux "ϕ" is defined as the number of mediators crossing unit area per unit time.

Now since the photons of light have been assumed to fall like matter, and since these particles can consist only of kinetic energy, it follows that the interaction "cross-section" for mediators must be proportional to total energy "E".

Hence the force on an elementary particle must be proportional to the product "$\phi.E$". Then to maintain symmetry it follows that the rate of production of mediators from any attractor must also be proportional to the total energy "E_s" of the attractor as well as proportional to "E". Noting that:-

$$m = \frac{E}{c^2}$$

the quantum-based replacement of the Newtonian expression becomes:-

$$F = \left(\frac{G}{c^4}\right) \cdot \frac{E_s \cdot E}{r^2} \qquad [1.4]$$

Newton gave an inverse square law based on the product of the masses i.e.;-

$$F = \frac{G \cdot m_s \cdot m}{r^2} \qquad [1.5]$$

Close inspection shows that two subtle changes have been incorporated as compared with the Newtonian physics. First it has already been shown that "c" is variable with respect to "r". But the gravitational force has to be proportional to "$E_s.E/r^2$", as already shown. Hence the so-called gravitational constant "G" is in fact a variable. The true constant is "G/c^4". It has the value:-

$$\frac{G}{c^4} = 8 \cdot 2615.10^{-45} \qquad N.m^2.J^{-2}$$

The second change is that the inverse square law only applies to the mediator flux, not the force. Since "E" reduces as "r" increases, the index to the law of force will exceed two. Now it is already known that pure non-precessing elliptical planetary orbits are only obtained when the index is exactly two. Trajectory computations show that a positive precession is obtained when the index exceeds two. An example of a half-orbit is given in FIG.12, showing what is meant by precession. Hence already a reason for precession is emerging. It remains to quantify the effect, but this is attempted

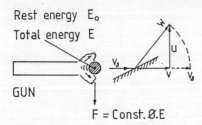

Rest energy E_0
Total energy E

GUN

$F = Const. \emptyset . E$

$$E = \frac{E_0}{\sqrt{1 - \left(\frac{w}{c_1}\right)^2 \Big/ \left(\frac{c}{c_1}\right)^2}}$$

FIG.3   MOTION IN AN ARBITRARY DIRECTION

much later.

1.2.5   MOTION IN ANY DIRECTION

Equation[1.1] was developed in zero gravitational field. The same result is returned if motion perpendicular to the field direction is assumed, normally known as the horizontal direction. It is necessary to provide a force which balances gravity. However, no work is done by either this force or that of gravity because motion in the vertical direction does not then take place. If now the supporting force is removed, so that the object now exists in a state of free-fall and is immediately bounced from an obstruction, the situation illustrated in FIG.3 will obtain.

The object rebounds elastically from a rigid plate and is immediately deflected without loss of speed. The final velocity "w", having an arbitrary direction, will have the same speed as "v_1", the initial horizontal velocity.

Also the derivation can be carried out at any level where the luminar speed is "c" as compared with the datum value "c_1".

In order to relate to a common datum for motion in any direction equation[1.1] is readily modified to the form:-

$$E = E_o \cdot \frac{1}{\sqrt{1 - \left(\dfrac{w}{c_1}\right)^2 + \left(\dfrac{c}{c_1}\right)^2}} \qquad [1.6]$$

1.2.6 RELATING LEVELS

Now it is necessary to find a way of relating all levels to one another. This is most simply achieved by imagining an object in a state of "resisted fall" as illustrated in FIG.4. This state can be defined as a form of motion in which a force is provided which exactly balances the force of gravity at all times. Motion of an object can then persist under its own inertia in any direction at an arbitrary speed, because it is not subjected to any net accelerating force. No energy can be stored by the object because work done by the equal and opposite forces exactly cancel one another. Hence the total energy of the object is not affected. If the object changes from level 1 to level 2, as shown in FIG.4, then the final total energy "E_2" will equal the initial value "E_1". A special case is the lifting of an object on a cable at a slow speed. Then total energy is equal to the rest value "E_o".

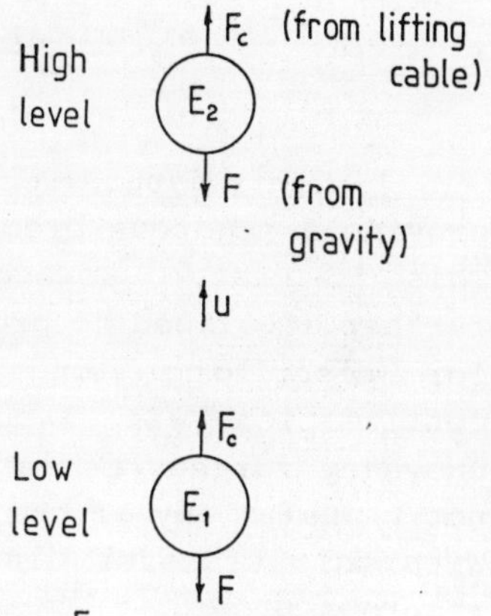

High level — E_2 — F_c (from lifting cable)

F (from gravity)

u

Low level — E_1 — F_c

F

$F_c = F$

$\therefore E_2 = E_1$ so $E_{o2} = E_{o1}$

Hence $m_{o2}.c_2^2 = m_{o1}.c_1^2$

$\therefore m_o = m_{o1}.c_1^2/c^2$

FIG.4 RESISTED FALL
or lifting an object on a cable

IT FOLLOWS THAT THE REST ENERGY "E_o" OF ANY OBJECT

CANNOT CHANGE NO MATTER HOW IT MOVES IN A GRAVITATIONAL FIELD.

Then this gives the relation needed to link all levels to one another. The significance of the deduction is that if in free fall the way "E" varies can be established, then the kinetic component "δE" will be known, because it is equal to:-

$$\delta E = E - E_o$$

The very astute reader may object that this does not provide a rigorous proof that rest energy is invariant. This is true. However later on, as the theory is further developed to provide a "negative buoyancy model" for describing long range forces, the necessary confirmation will appear. Briefly, in producing buoyancy forces, mediators are not absorbed and so do not transfer any of the energy from which they are made. With this transfer disallowed the above *does* become a rigorous proof.

This type of force contrasts with the absorption models needed to explain the nuclear strong and weak forces for which the above argument is inapplicable. The way binding energies relate to observed masses for these short-range forces cannot therefore be extended to relate to those of gravitation. A similar situation will apply to the remaining long-range force of electromagnetism.

Furthermore, since "c" changes with level in the gravitational field and "$E_o = m_o.c^2$", it follows that the rest mass "m_o" must change with level. This compensates for the variation in "G" with level, i.e.:-

$$m_o = m_{o1}.\left(\frac{c_1}{c}\right)^2 \qquad [1.7]$$

It also follows that "inertial mass" is a property secondary to energy. Mass allows accelerations to be calculated from impressed forces but does not represent the fundamental "substance" from which matter is constructed.

*Energy, therefore, needs to be regarded as the
primary building substance of both matter and motion.*

It will also be appreciated that the force of
gravity is also not mass-dependent as assumed by Newton.
At lower levels in a uniform gravitational field a given
value of mass will have a smaller weight than at a
higher level, even though its inertia remains the same.
Only for a given total energy will the weight remain
constant with level. Then the inertia of the object
concerned will increase as level reduces. These are
important new predictions of the extended Newtonian
approach.

It is also worth noting at this point that a
parallel exists with a charged particle in an electric
field. The same argument can be used to show that "E_o"
will remain constant with "E" rising as the particle
gains speed when free falling by electrical attraction.
In this case, however, "c" remains fixed and so the rest
mass now also remains constant. Also, because electric
charge remains constant an exact inverse square law will
now apply if the same mediator flux model is used by way
of explanation.

It follows that an object will have a rest energy
equal to the sum of the rest energies of its components,
when bound by either gravitational or electric forces.
It is assumed that in either case the components are
drawn in from an initially infinite distance and have no
motion in the assembled state. If, however, they are in
motion, such as electrons in the orbitals of atoms, then
the rest energy of the assembly will exceed the total of
its components. It will not be possible to relate
binding energies to masses by direct weighing as in the
case of the binding energies of nuclear forces, because
in these cases the assembled masses are always less than
that of the components, the differences being the
binding energies.

The new deduction runs counter to accepted beliefs.

The reason for the difference between these long-range forces and the short-range ones involved in the strong and weak nuclear forces is explained in Chapter 5 of the non-technical section.

1.2.7 GRAVITATIONAL POTENTIAL

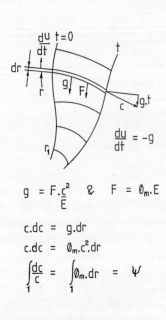

$$g = F.\frac{c^2}{E} \quad \& \quad F = \emptyset_m.E$$

$$c.dc = g.dr$$

$$c.dc = \emptyset_m.c^2.dr$$

$$\int_1 \frac{dc}{c} = \int_1 \emptyset_m.dr = \psi$$

FIG.5   DEFINITION OF ψ

ψ - Gravitational
Potential

The position has now been reached at which an expression defining the "gravitational potential" denoted by "ψ" can be derived. A vertical array of light-beams of the kind considered in FIG.1 is to be imagined. This is illustrated in FIG.5 for the general case in which a multiplicity of arbitrary attractors act to create the field. Because all beam elements started out perpendicular to the field direction and provided only a short time "t" is allowed to elapse before they reach the left boundary, the energy change in the field direction can be made indefinitely small. The simplification of Newton's law to the vertical acceleration being directly proportional to force divided by mass is then justifiable. The intention is to carry out an integration in the field direction to find an expression giving the end-to-end values of luminar speed.

By combining previous expressions [1.3] and [1.4]

for the gravitational force, noting that "F = g.m", identical with:-

$$F = g \cdot \frac{E}{c^2}$$

and defining flux with the constant of proportionality combined as "ϕ_m", then "F" is also given by:-

$$F = \phi_m \cdot E$$

Equating we can write:-

$$\phi_m \cdot E = g \cdot \frac{E}{c^2}$$

And $$g = c \cdot \frac{dc}{dr}$$

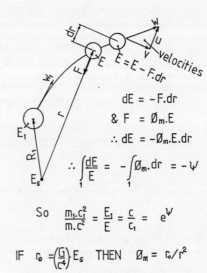

$$dE = -F.dr$$
$$\& \ F = \phi_m.E$$
$$\therefore \ dE = -\phi_m.E.dr$$
$$\therefore \int_1 \frac{dE}{E} = -\int_1 \phi_m.dr = -\psi$$

So $$\frac{m_1.c_1^2}{m.c^2} = \frac{E_1}{E} = \frac{c}{c_1} = e^\psi$$

IF $$r_0 = \left(\frac{G}{c^4}\right) E_s$$ THEN $$\phi_m = r_0/r^2$$

So $$\psi = \frac{r_0}{R_1}\left(1 - \frac{R_1}{r}\right)$$

FIG.6   FREE FALL

Hence:-

$$\phi_m \cdot E = \left(c \cdot \frac{dc}{dr}\right) \cdot \frac{E}{c^2}$$

Then simplifying we have:-

$$\int_1 \frac{dc}{c} = \int_1 \phi_m \cdot dr = \psi \qquad [1.8]$$

This is the required definition of gravitational potential "ψ".

The corresponding total energy variation for an object in free fall is also required. The free fall situation is illustrated in FIG.6. The total energy variation in the field can be found by equating the work done against gravity, "-F.dr", to the energy change "dE" in free fall or rise. Hence:-

$$dE \;=\; -F.dr \;=\; \phi_m.E.dr$$

Which can be rearranged to yield:-

$$\int_1 \frac{dE}{E} \;=\; -\int_1 \phi_m.dr \;=\; -\psi \qquad [1.9]$$

Equations [1.2], [1.8] and [1.9] can be integrated and combined to yield:-

$$\frac{c}{c_1} \;=\; \frac{E_1}{E} \;=\; \frac{m_1}{m}.\left(\frac{c_1}{c}\right)^2 \;=\; e^{\psi} \qquad [1.10]$$

For the special case of a single attractor the force given by equation[1.4] can be equated to "$\phi_m.E$" so that "ψ" becomes:-

$$\psi \;=\; \int_1 \phi_m.dr \;=\; \frac{G}{c^4}.E_s.\int_1 \frac{dr}{r^2}$$

$$or:- \qquad \psi \;=\; \frac{G}{c^4}.E_s.\left(\frac{1}{r_1} - \frac{1}{r}\right)$$

If now the "gravitational radius" "r_o" is defined as:-

$$r_o \;=\; \frac{G}{c^4}.E_s \qquad [1.11]$$

then the gravitational potential becomes:-

$$\psi \;=\; \frac{r_o}{r_1}.\left(1 - \frac{r_1}{r}\right) \qquad [1.12]$$

Already a stage has been reached at which some very useful deductions can be drawn. As "r" is reduced below "r_1", equation[1.12] shows "ψ" to become negative. Then equation[1.10] shows the value of "c" to fall below "c_1". But "c" only falls to zero when "r" becomes zero. This means that for q u a n t u m - b a s e d gravitation no "Black Hole" can exist.

In general relativity the speed of light falls to zero at the s o - c a l l e d "Schwarzchild radius" given by "$2.G.m_s/c^2$". This defines the "event horizon" of the Black Hole. Inside, the speed still falls, yet it cannot because it is already zero. Hence, a breakdown in logic has arisen. Also, if Tryon's(123) figure for the average mass

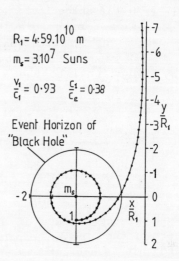

$$R_1 = 4.59.10^{10} \text{ m}$$
$$m_s = 3.10^7 \text{ Suns}$$
$$\frac{v_1}{c_1} = 0.93 \quad \frac{c_1}{c_e} = 0.38$$

Event Horizon of "Black Hole"

FIG. 7 TRAJECTORY STARTING INSIDE A "BLACK HOLE"

density of the universe he gives as 8.10^{-27} kg/m^3 is used together with an assumed radius for the universe of 10 billion light years, then general relativity predicts a black hole several billion light years across. Not even light can emerge from a Black Hole. Hence the universe could never have emerged! The new theory makes the "BIG BANG" of creation more plausible because no such problem now exists.

To illustrate the advance the new theory makes the trajectory of an object starting from *inside* a black hole is shown in FIG.7. The theory of curved trajectories on which this computation depended is given later. It is interesting to note that in such strong fields the speed is actually *increasing* as total energy reduces. This is due to "c" increasing rapidly.

1.2.8 THE GRAVITATIONAL RED SHIFT

At this stage it is possible also to obtain a simple expression for the gravitational time dilation effect known as the "gravitational red shift".

Max Planck showed that the energy "E" of a photon is proportional to the associated wave frequency "ν", indeed:-

$$E = h.\nu \qquad\qquad [1.13]$$

Used with equation[1.10] the result is:-

$$\frac{\nu}{\nu_1} = \frac{E}{E_1} = e^{-\psi}$$

Even at the surface of a massive star only the first term of the binomial expansion for EXP($-\psi$) need be taken and this is equal to "$1 - \psi$". Also the small change in frequency "$\delta\nu/\nu$" is:-

$$\frac{\delta\nu}{\nu} = (1 - \psi) - 1 = -\psi$$

From equation[1.12] the result can be simplified to:-

$$\frac{\delta\nu}{\nu} = \delta\left(\frac{G.m_s}{c_1^2.r}\right) = \delta\left(\frac{G.E_s}{c_1^4.r}\right) \quad [1.14]$$

This is exactly the same as the expression for the gravitational red shift taken from Chiu and Hoffmann(203) and derived by general relativity. Will(124) describes how this equation has been verified by experiment. The way this has now to be interpreted is that a photon loses energy when rising against gravity, hence its frequency is reduced at the higher level, so shifting a spectral line toward the red end of the spectrum. The observer at high level receiving such light therefore interprets vibrations to be running slower at a lower level.

Another way of finding gravitational time dilation can imagine a pair of masses vibrating on a spring of a given stiffness. The vibrating system is lowered on a cable. The masses increase as level falls since their rest energy remains constant and so the frequency should be smaller at a lower level.

Two equal masses "m_o" are in vibration due to connection by a spring of constant stiffness "$2.k_s$", so that the stiffness of the half-spring is "k_s". Then the frequency "ω" at any level will be "$\sqrt{(k_s/m_o)}$". Now equation[1.7] can be used to relate the mass to the datum level where it will be "m_{o1}". Then using equation[1.10] also the result can be written:-

$$\omega = \sqrt{\frac{k_s}{m_o}} = \frac{c}{c_1}.\sqrt{\frac{k_s}{m_{o1}}} = e^{\psi}\sqrt{\frac{k_s}{m_{o1}}}$$

i.e. $$\frac{\omega}{\omega_1} = e^{\psi} \sim 1 + \psi$$

Again the frequency is observed to be lower when the object is viewed from a higher level. The apparent contradiction is readily seen on close inspection to be due to the difference in the way the two systems operate but the end result is identical for both. Both methods are consistent with one another.

The theory needs further development, however, before other comparisons with the predictions of general relativity can be attempted. The modifications will not affect conclusions already drawn, as will be shown.

1.2.9 MASS AND ANGULAR MOMENTUM

Since "c" varies with "ψ" and the rest energy "E_o" does not change, and since "$E_o = m_o.c^2$", it follows that "m_o", the "rest-mass" of an object must vary with "ψ". Indeed from equations[1.7] & [1.10] it follows that:-

$$m_o = m_{o1}.\left(\frac{c_1}{c}\right)^2 = m_{o1}.e^{-2.\psi} \quad [1.15]$$

But the total mass in free fall can be obtained by further substitution from equation[1.10] as:-

$$m \;=\; m_1 \cdot \left(\frac{c_1}{c}\right)^3 \;=\; m_1 \cdot e^{-3 \cdot \psi} \qquad [1.16]$$

Angular momentum needs to take account of the change in "m" with "ψ" and so instead of the product "v.r" remaining constant in the gravitational field, "m.v.r" remains constant so that:-

$$v \cdot r \;=\; v_1 \cdot r_1 \cdot \frac{m_1}{m} \;=\; v_1 \cdot r_1 \cdot e^{3 \cdot \psi}$$

$$\dots\dots[1.17]$$

The virtual mass of space will be similarly affected and, as will presently be shown, implies that space is compressible.

1.3.0 THE COMPRESSIBILITY OF SPACE.

Since the present theory is quantum-based, it is perfectly reasonable to integrate with it other equations arising from quantum mechanics. Novikov(211) describes the virtual particles of space as occupying tiny hypothetical cubes of side "L". Their size is so chosen that one new virtual particle arises on average within each unit cube, just as its predecessor expires, again at some point within the same cube. An attempt to visualise this is made in FIG.8. The picture is complicated by the virtual particles having a whole range of masses, because "L" varies with "m", according to Novikov, by the expression:-

$$L \;=\; \frac{\hbar}{m \cdot c} \qquad [1.18]$$

It may be objected that since the existing quantum theory has been modified, deductions from it cannot be justified. What is indicated, however, is the existence of a parallel solution. If equation[1.18] is only assumed to be valid for the time being, it will later be found consistent with the new approach.

It is reasonable to assume that the ratio of total mass "m" to rest mass will be the same on average for all virtual particles arising throughout space. With "m/m_o" constant the situation is equivalent to resisted fall, so that from equation[1.15] the ratio "m/m_1" is equal to "EXP$(-2.\psi)$".

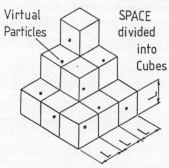

Virtual Particles

SPACE divided into Cubes

Novikov $\quad L = \dfrac{\hbar}{m.c}$

So $\quad \dfrac{L}{L_1} = \dfrac{m_1.c_1}{m.c}$ But $\dfrac{m_1}{m} = \dfrac{c^2}{c_1^2}$

$\therefore \quad \dfrac{L}{L_1} = \dfrac{c}{c_1} = \dfrac{E_1}{E} = e^{\psi}$

FIG.8 NOVIKOV'S SPACE CUBES
Shows space to be compressible

There will be some average value of "L" corresponding to the average value of "m", taking account of the wide distribution of virtual mass of the particles of space at any point. The symbol "L" refers to this average henceforth. So the ratio of "L" at some value of "ψ" will be related to the value "L_1" at "ψ" = 0. From equations[1.15] the above & [1.10] it follows that:-

$$\frac{L}{L_1} = \frac{m_1 . c_1}{m . c} = \frac{c_1}{c} . e^{2.\psi} = \frac{e^{2.\psi}}{e^{\psi}}$$

It follows that:-

$$\frac{L}{L_1} = \frac{c}{c_1} = \frac{E_1}{E} = e^{\psi} \qquad [1.19]$$

Since this implies that the virtual particles of space are packed more closely together at a lower level in the field, space can be considered as compressed by the proximity of ponderous matter. The situation is

very much the same as the compression of a gas by the
gravitational field. It is interesting to note that the
reduction of spacing has the same ratio as that of
light. This is totally different from Einstein's
"curved space-time", because the initial assumptions are
different.

1.3.1 THE TRUE SPEED OF LIGHT

The term "TRUE SPACE" will be reserved to mean the
case in which the compressibility of space is taken into
account. The term "FALSE SPACE" can then apply to the
case of incompressible space. In order to appreciate
the effect of compressibility a quantum description of
motion needs to be incorporated.

The description to be summarised is consistent with
the phenomenon of "wave-particle duality". However the
"Copenhagen interpretation" and the "Everett-Wheeler"
hypothesis are both inconsistent with the present
approach to quantum gravitation. A new solution is
described in detail in the non-technical part of this
book which fits the new theory to perfection.

Instead of a photon of light moving in a straight
line, like a rifle bullet, it is to be imagined as
moving in a series of jerks. The model is illustrated
in FIG.9. Photon energy is assumed to propagate at a
speed so high as to be infinite in effect. But because
space is filled with virtual particles and a substantial
proportion are electrically charged, and since photons
couple strongly with charge, they cannot penetrate very
far. They are absorbed by a virtual charge after moving
an average distance "L_p" which is proportional to "L".
The photon stays absorbed for the remaining life of the
particle and is then re-emitted to jump to the next
absorber.

1.3.2 A STRUCTURE FOR THE ELECTROMAGNETIC WAVE

This model has been extended to provide a physical
structure able to describe propagation of the
electromagnetic wave. It answers the question of how
electromagnetism can be propagated by the electrically

neutral photon. This is more fully treated in the technical book(212) but will be briefly described here.

FIG.6 in Chapter 10 shows a spin coupling scheme for mediators to describe the electric force. A static vertical electric field is represented by a pair of charged spheres in which a positive charge is placed vertically above a negative charge. The forces acting on both an electron and a positron lying on the horizontal mid-plane are also shown. The triangles of force, caused by the selective action of

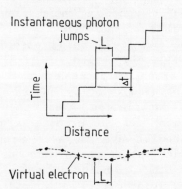

Structured Electromagnetic Wave

$$c_1 = \frac{L_1}{\Delta t_1} \qquad c = \frac{L_1}{\Delta t} \qquad c_T = \frac{L}{\Delta t}$$

$$\frac{\Delta t_1}{\Delta t} = \frac{c}{c_1} = \frac{L}{L_1} \quad \text{So} \quad \frac{c_T}{c_1} = \frac{c^2}{c_1^2} = e^{2V}$$

$$\text{YIELDS} \quad \frac{dc_T}{dr} = 2 \cdot \frac{dc}{dr}$$

FIG. 9   TRUE SPEED OF LIGHT c_T

mediators, indicate that net accelerations will be vertical.

Now it is to be imagined that the charges on the spheres are cycling instead of remaining static. One half cycle later the charges will be of opposite sign but of the same magnitude. Now the coupling by virtual mediators will create opposed vertical sinusoidal motions of the electron and positron but because their charges are opposite, the effects are additive. The phase lag of this motion as compared with that of the source will increase progressively with distance. When a large number of virtual particles are considered stretching out in a long horizontal line, the net effect will be to generate an electric transverse wave propagating at the speed of light. The virtual charged

particles are a small fraction of the total which
inhabit space. Their fleeting lives complicate the
model but the overall effect remains the same, the
electric field being associated with this transverse
oscillatory motion.

Each particle of the chain reaches its maximum
transverse velocity whenever it crosses the mid point of
its oscillation. But they are charged particles in
motion and so produce a magnetic field. Hence a
magnetic wave is also propagated, though a quarter
period out of phase with the electric wave.

All textbooks show these waves in step but this is
clearly an error. Electric and magnetic energy have to
propagate such that the sum of the two is constant. If
they were in phase the energy would keep rising to twice
the average and fall to zero twice each cycle. The
pattern also needs to follow that of the source in which
there will be zero current, and therefore magnetic
field, at the point when the spheres are at peak
electric potential.

The electric field represents potential energy, the
magnetic field the kinetic form. The two forms keep
interchanging to create a travelling resonant system and
the amplitude at any point is determined by the energy
density carried at that point.

The energy of the wave cannot be carried by the
mediators causing the transverse motion, because these
alternately supply positive and negative energy in equal
and, therefore, cancelling amounts. Also, as will be
shown in Chapter T.S.3, such mediators cannot in any
case be absorbed; they act by creating a buoyancy type
of force produced by a partial density gradient. Hence
mediators of the electric force do not act in the same
way as photons. The energy transport needs to be
carried by real photons which are, however, outnumbered
by mediators by many thousands to one. Hence the
waveform is unaltered by the real photons which
temporarily energise the charged particles of space in
a sequential manner. It is unlikely that the photon
ever exists as a separate particle because its jumps

imply almost infinite propagation speed.

According to the present theory, energy can only propagate so fast when travelling along filaments of the Grid. The model is therefore consistent with photon energy being transmitted along the filaments to be incorporated in the next suitable charged particle undergoing re-creation. This will also throw off mediators travelling at the speed of light during its brief life according to the scheme given FIG.6[*] of Chapter 10. Its brief life will terminate to provide the average dwell time "δt". This is deliberately programmed by the Grid to increase as level falls in order to provide a consistent system of mechanics.

We are now in a position to recontinue development of that part of the theory directly affecting gravitation. The dwell-time "δt" of a photon at each absorber divided into the distance "L_p", then gives the true luminar speed "$c_T = L_p/\delta t$". The false luminar speed at the same level would be "$c = L_{p1}/\delta t$". The time delay "δt" is the same for both cases. Hence making use of equation [1.19] we can write:-

$$\frac{c_T}{c_1} = \frac{c}{c_1}\cdot\frac{c_T}{c} = \frac{c}{c_1}\cdot\frac{\dfrac{L_p}{\delta t}}{\dfrac{L_{p1}}{\delta t}} = \left(\frac{c}{c_1}\right)^2$$

Hence from equations [1.1] and [1.19]:-

$$\frac{c_T}{c_1} = \left(\frac{c}{c_1}\right)^2 = e^{2\cdot\psi} = \left(\frac{E_1}{E}\right)^2 = \frac{1 - \left(\dfrac{w}{c_T}\right)^2}{1 - \left(\dfrac{w_1}{c_1}\right)^2}$$

$$\dots\dots\dots\dots[1.20]$$

The equations [1.19] and [1.20] used in conjunction form the basic starting point for computing trajectories according to the present theory. The last term of

* See Page 178

[1.20] comes from "$(E_1/E)^2$" using [1.1]. It needs to be observed that "$c_{T1} = c_1$" because these are the datum luminar speeds. Also "w/c_T" not "w/c" appears since only when "$w = c_T$" can "E" become infinite. Only at apogee and perigee can "v" replace "w" since only then is "u" zero.

If equation[1.20] is differentiated with respect to "r" and compared with a similar result for "c/c_1", the two compare as follows:-

TRUE SPACE   FALSE SPACE

$$\frac{d\left(\frac{c_T}{c_1}\right)}{dr} = 2 \cdot e^{2 \cdot \psi} \cdot \frac{d\psi}{dr} : \qquad \frac{d\left(\frac{c}{c_1}\right)}{dr} = e^{\psi} \cdot \frac{d\psi}{dr}$$

Now in weak fields the exponential term is negligibly different from unity. It follows that the true change in speed of light with "r" is twice the value for false space.

1.3.3 THE TRUE GRAVITATIONAL BENDING OF A LIGHT-BEAM

Now the false space result corresponds with the Newtonian physics already discussed with reference to FIG.1.* By treating a real case as a succession of circular arcs joined end to end, the actual deflection of light by the Sun could be calculated. This is proportional to the change in speed of light with respect to distance in the field direction. Hence with the true change being twice that for false space, it follows that the observed deflection will be twice that predicted from the Newtonian. This is exactly the same as Einstein's prediction based on curved space-time and is supported by observation as described by Will(124).

1.3.4 THE SHAPIRO TIME DELAY

Also according to Will(124) Shapiro measured the time delay caused by sending a radar beam close to the Sun to be returned by a transponder fixed to the surface of Mars. Mars was close to "superior conjunction". The

* See Page 253

situation is illustrated in FIG.10. The beam is shown as a straight line because the bending effect is very much second order and can be neglected. The dots represent intervals between virtual particles, greatly exaggerated, to suggest the effect of space-compressibility.

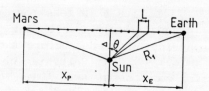

$$\delta t = \frac{4 \cdot r_0}{c} \left\{ \ln \left(\frac{4 \cdot |x_P \cdot x_E|}{\Delta^2} \right) - \frac{|x_E| + |x_P|}{R_1} \right\}$$

For $\Delta = 3.5$ Sun radii

	$\delta t \mu s$
Shapiro	131·7
Møller	151·4
Pearson	127·7

FIG.10   THE SHAPIRO TIME DELAY

The time delay is defined as the difference between the actual time taken for the double journey and the time which would have been taken in the absence of slowing of light by the gravitational field. From the theory given, relating "c_T/c_1" to "r" and using FIG.10, it is not difficult to deduce the equation for the time delay "δt" reproduced in FIG.10.

The first term is identical with that given by both Shapiro(214) and Møller(210). However there are differences in the secondary term. The differences between Møller and Shapiro are far greater than between Shapiro and the quantum derivation. In addition the equation Shapiro gives is not symmetrical, so that it would predict a different answer if the test were made by sending the signal from Mars instead of Earth. The difference is several orders of magnitude greater than could be accounted for by the difference between Earth and Mars gravitational potential. Unfortunately Shapiro does not give his derivation. However, as can be seen from the comparisons given in FIG.10, the predicted values of Shapiro and the quantum base are too close for discrimination by experimental observation.

1.3.5 PRECESSION OF PLANETARY ORBITS

A further development in the extended Newtonian physics, is needed before motions in curved paths can be calculated. If Cartesian coordinates are adopted, then for motion in two dimensions mutually perpendicular forces "F_x" and "F_y" need to be considered, acting simultaneously. It is first necessary to determine equations of motion for speeds up to that of light without resorting to relativity.

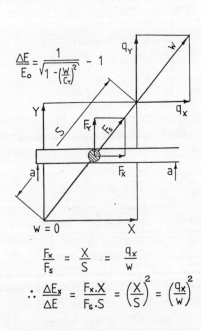

$$\frac{\Delta E}{E_o} = \frac{1}{\sqrt{1 - \left(\frac{w}{c_T}\right)^2}} - 1$$

$$\frac{F_x}{F_s} = \frac{X}{S} = \frac{q_x}{w}$$

$$\therefore \frac{\Delta E_x}{\Delta E} = \frac{F_x \cdot X}{F_s \cdot S} = \left(\frac{X}{S}\right)^2 = \left(\frac{q_x}{w}\right)^2$$

FIG. 11 ENERGY COMPONENTS

These can be readily obtained by considering a projectile fired by a gun pointing in the "X" direction. At the same time the gun-barrel is given acceleration "a" in the "Y" direction and care is taken to ensure that at all times the direction in which the gun points remains fixed. This acceleration "a" is so controlled that the projectile moves in a straight line as seen by a stationary observer. This can only be achieved by making the ratio of forces acting on the projectile "F_x/F_y" constant at all times.

The result is depicted in FIG.11. It is clear that three similar triangles exist so that the distances moved, the forces involved and the terminal velocity components reached bear a simple relation to one another. If "S" is the absolute distance moved and "w" is the terminal velocity, whilst "q" represents velocity components, it follows that:-

$$\frac{F_x}{F_s} = \frac{X}{S} = \frac{q_x}{w} \quad \& \quad \frac{F_y}{F_s} = \frac{Y}{S} = \frac{q_y}{w}$$

But $\quad \delta E_x = F_x . X \quad \& \quad \delta E = F_s . S$

The latter is the total kinetic energy. Hence:-

$$\frac{\delta E_x}{\delta E} = \frac{F_x . X}{F_s . S} = \left(\frac{X}{S}\right)^2$$

So:- $\quad \dfrac{\delta E_x}{\delta E} = \left(\dfrac{q_x}{w}\right)^2 \quad \& \quad \dfrac{\delta E_y}{\delta E} = \left(\dfrac{q_y}{w}\right)^2$

$$\dots\dots[1.21]$$

Now "$\delta E = E - E_o$" and so from equations [1.6] and the logic of [1.20] it follows that what might be termed the "directed energy component" in the "X" direction is given by:-

$$\frac{\delta E_x}{E_o} = \left[\frac{1}{\sqrt{1 - \left(\dfrac{w}{c_T}\right)^2}} - 1\right] . \left(\frac{q_x}{w}\right)^2$$

$$\dots\dots\dots[1.22]$$

Where $\quad \dfrac{w}{c_T} = \dfrac{w}{c_1} . \dfrac{c_1}{c_T} = \dfrac{w}{c_1} . e^{-2 . \psi}$

A similar expression can be written for "$\delta E_y / E_o$" and by an extension to three dimensions a similar expression for "$\delta E_z / E_o$" results. All the basic mechanics required for the computation of planetary orbits has now been

provided.   It does, however, require a considerable
amount of mathematical processing in order to transform
these expressions into useable equations of motion.   A
full treatment is outside the scope of this lecture.

However, it is first necessary to differentiate
equation[1.22] with respect to time.   Then:-

$$\frac{d(\delta E_x)}{dt} = \frac{d(F_x \cdot X)}{dt} \quad \& \quad \frac{dx}{dt} = q_x$$

For the special case of a single massive attractor
placed at the origin "F_x" is related to "F" by:-

$$F_x = \frac{F \cdot X}{\sqrt{X^2 + Y^2}}$$

By this step useable equations of motion result.

It has been found
more   appropriate   to
work in polars rather
than   in   Cartesian
coordinates.   In this
case "u" and "v" are
used instead of "q_x"
a n d   " q_y "   i n
equation[1.22].
Computations   using
exact equations written
in polars showed that
precession   steadily
increased   as   the
a t t r a c t o r   m a s s
increased.   More
surprising was the
progressive   dis-
conservation of angular
momentum,   even using
t h e   r e v i s e d
expression[1.17].   The
latter is applicable
only to the case of

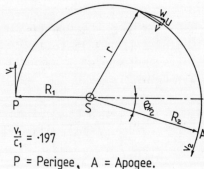

$\frac{v_1}{c_1} = \cdot 197$

P = Perigee,   A = Apogee.

S = 10^6 Suns, R_1 = 45·9 Mkm.

Half orbit shown. R_1 & R_2 would form
a straight line for a pure ellipse.

FIG.12 PRECESSION OF ORBIT
c_1 is at datum point P .
β is the angle of precession.

false space. For both false and true space, however,
the same progressive dis-conservation of angular
momentum was predicted. In each case extrapolation back
to zero velocity showed that no dis-conservation would
arise in very weak fields. It was therefore judged
reasonable to use a simpler method of analysis in which
angular momentum was assumed conserved in weak fields.
The method used will now be outlined.

1.3.6 ANGULAR MOMENTUM FOR TRUE SPACE

The mass variation with gravitational potential will
be the same as for false space, since this depends on
"c/c_1" not "c_T/c_1". But the law of force remains the same
for both true and false space because an inverse square
law of mediator flux variation applies equally. Hence
"E/E_o" is the same for both cases as given by
equation[1.1] at any given radius from an attractor.
Hence "v_T/c_T" can replace "v/c". So "$v_T/v = c_T/c$" and
will therefore vary as "L/L_1". With "m/m_1" from
equation[1.15] and "L/L_1" from[1.19] it follows that:-

$$v.r = v_1.r_1.e^{4.\psi} \qquad [1.23]$$

This is equivalent to saying that the quantity which
should be most nearly conserved is the product
"$m.v.r/L$". Computations have shown that exactly the
same orbital shapes, dis-conservation of angular
momentum and orbital precession values are obtained as
those computed for false space. Only the velocities are
modified. This is exactly in accord with the
expectations from simple logic, because all velocity
components are affected in the same ratio by the change
from false to true space.

Computations made using exact equations and
extrapolated back to zero field showed that angular
momentum would be conserved at this limiting condition.
This law makes for simpler computation.

Equations which describe the mechanics on this basis
are readily formulated. A start can be made by
assembling previous equations to yield:-

$$\left(\frac{L}{L_1}\right)^2 = \left(\frac{c}{c_1}\right)^2 = \frac{c_T}{c_1} = \left(\frac{E_1}{E}\right)^2 = e^{2 \cdot \psi}$$

$$\dots \dots [1.24]$$

Then with [1.20] and noting that "$c_{T1} = c_1$" above yields an expression giving "w" i.e.:-

$$\left(\frac{w}{c_1}\right)^2 = \left[1 - \left[1 - \left(\frac{w_1}{c_1}\right)^2\right] . e^{2 \cdot \psi}\right] . e^{4 \cdot \psi}$$

$$\dots \dots [1.25]$$

$$and \qquad w^2 = u^2 + v^2 \qquad [1.26]$$

Now since "u" is zero at apogee and perigee, then at the perigee, used as datum, "$v_1 = w_1$", and at the apogee "$v_2 = w_2$". Then using equation [1.23] with [1.25] it is readily shown that:-

$$\left(\frac{v_1}{c_1}\right)^2 = \frac{1 - e^{-2 \cdot \psi_2}}{1 - \left(\frac{r_1}{r_2}\right)^2 . e^{2 \cdot \psi_2}} \qquad [1.27]$$

$$with \quad v = r . \frac{d\theta}{dt} \quad \& \quad u = \frac{dr}{dt} \qquad [1.28]$$

Equations [1.23] to [1.28] then provide a complete description of orbital motion. For the gravitational fields and with perigee and apogee radii taken equal to that of the planet Mercury, it was found that computational error made results unreliable below attractor masses less than about a million solar masses. A half-orbit corresponding with this case, obtained by assuming that both energy and angular momentum are conserved, is shown in FIG.12.

In order to investigate the weak fields available to observation it was therefore necessary to use the Newtonian ellipse as a base (suffix "N"). Then small differences in radial and tangential velocities i.e. "$\delta u = u - u_N$" and "$\delta v = v - v_N$" respectively could be found using expansions in series. Very simple expressions can be derived to give the tangential and radial velocity components for the Newtonian case. They are:-

$$v_N = v_{N1} \cdot \frac{r_1}{r}$$

$$\left(\frac{u_N}{v_{N1}}\right)^2 = \left(1 - \frac{r_1}{r}\right) \cdot \left(\frac{r_1}{r} - \frac{r_1}{r_2}\right)$$

$$\left(\frac{v_{N1}}{c_1}\right)^2 = \frac{2 \cdot \dfrac{r_o}{r_1}}{1 + \dfrac{r_1}{r_2}} \qquad [1.29]$$

The values of "δu" and "δv" obtained between the two cases can be integrated separately for a half Newtonian orbit and then used with the following expression, derived by simple logic, to determine the half-precession per orbit "$\beta/2$" given in radians by:-

$$\frac{\beta}{2} = \int_0^\pi \left(\frac{\delta v}{v} - \frac{\delta u}{u}\right) \cdot dQ \qquad [1.30]$$

Here "dQ" is an element of angle turned through in the Newtonian base trajectory.

It requires a considerable amount of tedious algebra to integrate and simplify the resulting expressions but this can be done, just as can those based on curved space-time. The resulting equations show that "$\delta v/v$" is generally positive with "$\delta u/u$" negative except at perigee and apogee where "$\delta u/u$" is zero. Hence both differences have an additive effect on the precession. The resulting equation for precession for any orbit is entirely different from that given by Einstein(110), however. The two are compared in FIG.13. In addition the results of calculations given by these two equations are compared and adequate data is also provided so that the reader can carry out an independent check.

It should be remembered that:-

1 arcsecond = radians.3600.180/π

The last two columns give the precession in arcseconds per orbit for Mercury, Venus, Earth and the asteroid Phaethon. Despite the totally different form of the equations they give almost identical results!

EINSTEIN $(r \equiv R)$

$$\beta = \frac{24.\pi^3.a^2}{T_0^2.c^2(1 - ecc^2)}$$

$$a = \frac{r_2 + r_1}{2} \qquad ecc = \frac{r_2 - r_1}{r_2 + r_1}$$

QUANTUM

$$\beta = 3.\pi.\left(1 + \frac{r_1}{r_2}\right).\frac{r_0}{r_1}$$

	r_1 Mkm	r_2	t days	β/orbit EIN$^{N.}$	QUA$^{M.}$
M	45·9	69·7	87·8	·1036	·1038
V	107·4	109·0	224·8	·0530	·0531
E	147·1	152·1	365·3	·0383	·0384
Ph	20·9	360·0	525·0	·1451	·1454

FIG.13   PRECESSION EVALUATIONS

It follows that general relativity must in some peculiar way be a mathematical parallel with extended Newtonian physics in which the conservation of angular momentum is assumed to apply. This is despite the totally different nature of the initial assumptions on which the two theories are based. But a difference arises in that for the curved space-time base the result is rigid and yet it is known to be slightly too high. This is the conclusion of Dicke, as reported by Will(108).

The Sun is not a perfect sphere; instead it is an oblate

spheroid. Dicke has worked out this ought to account
for about 1 arcsecond of perihelion advance per century,
leaving 42 arcseconds for the relativistic component,
instead of the 43 to which the values in FIG.13 yield
when converted to the century time-scale.

This is where the new theory offers a potential
advance. There must be a minute residual dis-
conservation of angular momentum because orbital speeds
are not zero although they are very small as compared
with light. This means that the result will be a slight
overestimate. Unfortunately the mathematical refinement
needed to verify this expectation has not yet been
carried out. It has proved very difficult indeed to
complete this final step but there is no fundamental
reason why it should not be possible.

Before proceeding to develop the theory further it
is pertinent to sum up the achievements made so far.

The equation for time dilation took no account of
the compressibility of space. However this is readily
shown to make no difference. This is because it is
connected with the way mass varies with gravitational
potential and this depends on the variation of "c", not
"c_T". In other words it depends only on the time delay
"δt" of photons between each absorption and re-emission
by virtual charged particles and this delay increases as
level falls so that "c" can fall. But the average
photon jumps across distances "L_p" are instantaneous and
so do not affect "m".

Hence the four most important experimental checks
which were responsible for establishing general
relativity can equally be used to support the new
quantum-based theory. In addition the null experiments
of Eötvös and Dicke, which have been interpreted as
supporting Einstein's principle of "equivalence", can
equally be interpreted as supporting the new theory,
because the coupling made here proportional to total
energy automatically satisfies these null experiments.
This is not all, however. The new theory predicts
gravitational waves. This aspect has not yet been fully

addressed unfortunately. Varying forces at a distance
have been predicted but the energy loss caused by wave
production has not yet been formulated. This ought to
yield a parallel solution to that obtained by Einstein
when somebody finds the correct lead to tackle the
problem.

In addition the new theory has been further
developed to show how space compression arises. Then
the way gravitation is linked to the electric force and
an explanation for the huge ratio of the two appears
quite readily.

1.4 RELATED PROBLEMS - NEGATIVE ENERGY STATES

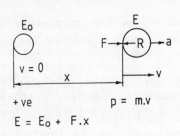

$E = E_0 + F.x$

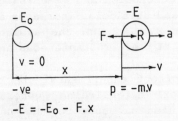

$-E = -E_0 - F.x$

FIG.14 EXTENDED NEWTONIAN
ACCELERATION

Chapter 5, making
reference to FIG.4[*] in
the non-technical
section, dealt with
negative energy states
in a manner which is
supplementary to this
section. They need to
be introduced at this
stage in order to
further develop the new
theory.

In FIG.14 the
upper system shows the
Newtonian assumption,
in which the force of
"action" "F" associated
with mechanical work
done **ON** an object
points in the same
direction as the
resulting acceleration.
The body itself opposes
this force by an equal
and opposite force of
"reaction" "R". It follows that the mass, energy and

momentum of the object all have to be considered
positive, because the work done "F.X" is positive.
Hence the kinetic energy gained, being equal to "F.X",
is positive. Then for "E" to exceed "E$_o$", as it must, to
be in accord with equation[1.1], both "E" and "E$_o$" also
need to be positive.

There is no way in which the initial assumption can
be checked, however. It is equally likely that the
forces of action and reaction "F" and "R" respectively
could be reversed.

In the lower figure such a reversal is shown.
Provided the signs of mass, energy and momentum are
reversed as well, exactly the same response occurs. This
is why it is impossible to be certain which kind of
system operates.

Now in FIG.2* a cone of mediators is depicted, from
which a quantum expression for the gravitational force
was deduced. If the mediators consisted of positive
mass/energy, then they would be positive momentum
carriers. They would produce a repulsion instead of an
attraction. It follows that these mediators must be
made of negative energy. The same inference applies to
the electric and nuclear forces. Hence to build a
universe both kinds of energy need to exist
simultaneously. The proton and the neutron must
themselves be composites of both kinds. If the net
energy of the three "quarks" from which they are thought
to be composed is positive, then the primary "gluons" by
which they are bound have to be negative.

Of course this is not accepted in present quantum
theory. Instead the artifice of the "negative coupling"
is used with carriers of positive momentum. But this is
an artificial way of reversing the force and has already
been shown invalid by the argument given in Chapter 5.

1.4.1 GETTING RID OF THE COSMOLOGICAL CONSTANT

Because both positive and negative states need to
exist, it is quite easy to arrange conditions so that
the net energy of the universe is zero. There are a

* See Page 254

number of ways of doing this but the only one which is
compatible with the new theory is illustrated in FIG.15.[**]
Space starts out with two huge components of positive
and negative energies in exact balance, so that the net
energy of space is zero. They are the "Yin" and the
"Yang" of energy relying on one another absolutely for
their   simultaneous   existence.   Expressed
mathematically:-

$$\varepsilon_+ - |\varepsilon_-| + \varepsilon_m = 0$$

where "ε_m"[*] is the energy density of all matter in the
universe. If in addition "ε_N" is the negative energy
superimposed on space, the "exnegmas" of the halo
causing gravitation, then:-

$$\varepsilon_N = \varepsilon_+ - |\varepsilon_-| = -\varepsilon_m \qquad [1.31]$$

The system exhibits instability, however with positive
matter tending to clump. This matter leaves a balancing
energy in the form of a tenuous halo extending for about
a billion light-years in all directions providing a
mediator flux density varying according to the inverse
square law. This arises because the halo consists of
virtual particles which are being continuously generated
at the surfaces of all elementary particles, only to
expire after a certain lifetime. This negative halo is
cross-hatched in the figure and exactly balances the
positive energy of the neutron star shown. Since the
halo consists of negative momentum carriers, it provides
a weak but universal attractive force on all positive
matter. It provides the primary cause of gravitation
responsible for the instability mentioned earlier.
Only about a quarter of all the gas in the universe is
assumed in FIG.15 to be in condensed form, so a
background of uniformly tenuous excess negative energy
will be present to balance this gas and is also
illustrated.

No unwanted "cosmological constant" is predicted.

In established theory with negative energy states
denied, the positive energy density of space has to be

[*] NOTE. $\varepsilon \equiv \mathbf{e}$ [**] See Page 288

balanced by the inferred "negative pressure of the
vacuum". The reasoning is condensed from Novikov(211).
Space is assumed to expand from zero volume by an amount
"δV" and the resulting energy gain equated to zero,
assuming that the vacuum can exert a pressure "P" i.e:-

$$\varepsilon.\delta V + P.\delta V = 0 \qquad [1.32]$$

Then Einstein's expression for energy density, "ρ"
which is "$(\varepsilon + 3.P)$", is assumed. It contains a
pressure term. But then because his equations do not
admit of a force, Newton's equation is utilised to give
the force on a particle sitting at the surface of a
sphere of radius "R" and of uniform energy density. The
combined result yields an acceleration "a" of:-

$$a = -\frac{4.\pi}{3}.(\varepsilon + 3.P).\frac{G.R}{c^2}$$

(i.e. an attractive force)
Then substituting the vacuum pressure "P" the result
yields:-

$$a = \frac{4.\pi}{3}.\frac{2.G.\varepsilon.R}{c^2} \qquad [1.33]$$

When "ε" is given the value 10^{45} J/m^3 and "R" is the
distance between galaxies, the value of "a" by this
equation requires a repulsive force about 10^{50} times
larger than astronomical observations can allow. If "a"
is divided by "R" then the "cosmological constant" is
obtained about which so much concern has arisen. The
current solution is to invent other universes in higher
dimensions communicating via "wormholes" in space to
produce cancelling effects, as described by Abbot(101).
The reader is asked to judge the plausibility of
this approach. In so doing the justification for
combining three theories, each based on a different
conception of reality, needs to be considered. The
degree of compatibility of the three sets of assumptions
needs to be taken into account. This needs then to be

weighed against the present approach in which because
negative energy states are allowed, only a very weak
force of attraction is predicted from a completely
consistent set of initial assumptions.

As far as we are concerned the problem of the non-
existent cosmological constant never arose!

1.5 GRAVITATION AS A NEGATIVE BUOYANCY FORCE

If the "exnegons", the particles of the halo, moved
in straight lines to be absorbed by particles of matter,
then the concentration gradient of the halo would also
have an inverse square law. It is then a simple matter
to determine "ε_{Ns}" the value of energy density of the
halo at the surface of a planet or star if the range of
the gravitational force is assumed. If this is assumed
to be a billion light years (10^9 years), then for the Sun
"ε_{Ns}" works out to -3110 J/m³. From this figure a
gravitational acceleration can be worked out and comes
out several orders of magnitude lower than the observed
value. This model is inconsistent with the buoyancy
model about to be described, however, and so cannot be
admitted.   The buoyancy model has to assume that
mediators *diffuse* through space. In this case, it can
be shown from equation[3.12] in Chapter T.S.3 that for
a range "r_m" the halo energy density at "r" is:-

$$\varepsilon_N \;=\; r_s \cdot \varepsilon_{Ns} \cdot \left(\frac{1}{r} - \frac{1}{r_m} \right)$$

This can be integrated and used with equation[1.31} to
yield the value of energy density "ε_{Ns}" at the radius "r_s"
of the surface of the attractor of mass "m_A" to give the
expression:-

$$\varepsilon_{Ns} \;=\; -\frac{3}{2 \cdot \pi} \cdot \frac{m_A \cdot c^2}{r_s \cdot r_m^2} \qquad [1.34]$$

The value for the Sun now works out at only
$-1.374 \cdot 10^{-12}$ J/m³! This is very low even compared with
the average balancing energy density of matter in the

universe which is only about $0.7.10^{-9}$ J/m^3.

So the primary mediators are far too diffuse for providing more than a tiny fraction of the observed force of gravity. They act on space as well as matter, however, and so provide the causative agent for the compression of space which has already been shown to exist.   The halo will pull the positive virtual particles of space together but repel the negative. Hence for a net compression to occur the interaction cross-section with the positive kind must exceed the cross-section for negative-negative interactions in the ratio "f_v". An identical imbalance also needs to apply to the sub-atomic particles of matter if the effect of space compression is to amplify the primary cause. The major part of the gravitational force is therefore seen to act in the manner of buoyancy.   Again FIG.15 illustrates this.

The two circles represent elementary particles acted upon by the pressure gradient associated with the energy density gradient describing compressibility. The upper one is exposed to only the positive half of space and so experiences a repulsion. The magnitude of this force is equal to the pressure gradient multiplied by the immersed volume just like any other buoyancy force. The lower is exposed to the negative half with similar but opposite results. It follows that again the negative mediators have to experience a greater volume than the positive ones in order that a net attraction arises. This could happen only if elementary particles are soft instead of hard, so that virtual positive particles penetrate further than negative ones during interactions.

This deduction leads to some very desirable predictions. The mediators flow round elementary particles without being absorbed. They are not used up as they would be if they were absorbed. The latter model would mean that the inner layers of a planet would be partially shielded so that gravitation would no longer be exactly proportional to the product of the total energies of the attracting objects.   During eclipses of the Sun a "gravitational moon shadow" would

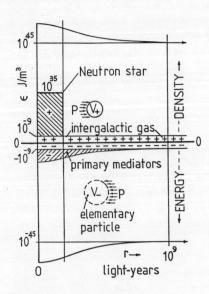

FIG. 15 SPACE and MATTER

Showing energies in
balance.

arise. These have been looked for but not the slightest effect has been found. This exactly accords with the buoyancy model. It remains to quantify the effect.

The new model needs to produce a force on each elementary particle exactly proportional to its total energy, since this was a primary deduction of the theory. Hence the energy density "ε_p" of all elementary particles needs to be a universal constant. The particles need to grow in volume as they acquire kinetic energy. This seems a very reasonable and satisfying result.

A third problem is resolved. If mediators are absorbed, then a tangential component of force will be induced on a planet in proportion to its orbital speed. The mediators stream out from a point source in a radial direction. The motion of the planet, however, induces a relative "wind". If they were positive a drag would arise, but being negative they act to provide a gain in angular momentum. The planet would spiral out of orbit. With buoyancy-type forces no tangential components arise.

It is now necessary to show that space compression can give about the right magnitude for the force of gravity.

Now the energy density of space "ε" will vary as the

inverse cube of "L" and so substitution from
equation[1.19] yields:-

$$\varepsilon = \varepsilon_1 \cdot \left(\frac{L}{L_1}\right)^{-3} = \varepsilon_1 \cdot e^{-3 \cdot \psi} \qquad [1.35]$$

Hence the energy density of space increases as "ψ"
reduces. An energy density gradient "$d\varepsilon/dr$" will
exist. This has an associated pressure gradient
"dP/dr" and acts on all elementary particles to produce
the gravitational force. On differentiating
equation[1.35] with respect to "r":-

$$\frac{d\varepsilon}{dr} = \varepsilon_1 \cdot (-3) \cdot e^{-3 \cdot \psi} \cdot d\frac{\psi}{dr} \qquad [1.36]$$

The mediators need to last on average about a
billion years to provide the observed minimum range.
The application of Heisenberg's uncertainty principle to
these virtual particles then shows that they need to
have zero rest energy. Consequently they will travel at
the speed of light and this will be the assumption made.
If mediators moved more slowly the logic to be developed
would not be significantly altered. It would simply
mean that the energy density of space needed would be
increased.

Also "j" is the number of mediators crossing unit
area per second, when local space contains a number
density "n" of mediators per unit volume. Each moves
with random orientation and has its direction "θ"
measured normal to the unit area.

With velocity vectors for mediators represented as
a uniform distribution with their arrow heads touching
the surface of a sphere having a radius represented by
distance "c", those within an element from the normal at
an angle between "θ" and "$\theta + \delta\theta$" will contain a number
"δn" out of the total number "n" in the ratio of surface
areas of this element as compared with the surface area
of a complete sphere and is:-

$$\frac{\delta n}{n} = \frac{2 \cdot \pi \cdot c \cdot \sin\theta \cdot c \cdot \delta\theta}{4 \cdot \pi \cdot c^2}$$

Each mediator will have a velocity component in the normal direction of "c.cosθ". Hence the number crossing unit area per second "j" becomes:-

$$j = \frac{n.c}{2} . \int_0^{\frac{\pi}{2}} \sin\theta.\cos\theta.d\theta = \frac{n.c}{4}$$

$$............[1.37]$$

And since the momentum carried per particle is "m.c.Cosθ" measured normal to the unit area, further integration gives the total momentum flux, which can be identified with pressure as:-

$$P = \frac{n.m.c^2}{6} \quad i.e. \quad P = \frac{\varepsilon}{6} \quad [1.38]$$

Then force caused by dP/dr acting across volume "V" is:-

$$F = V.\frac{dP}{dr} = \frac{V}{6}.\frac{d\varepsilon}{dr} \qquad [1.39]$$

1.5.1 THE MECHANISM OF SPACE COMPRESSION

A virtual particle of space will have an energy "E_v,", will occupy a volume "V_v" and will consequently have an energy density "ε_v", so that:-

$$\varepsilon_v = E_v/V_v$$

The buoyancy force acting on it due to the exnegmas, (suffix "x") alone will be:-

$$F = -\frac{V_v}{6}.\frac{d\varepsilon}{dr} = -\frac{1}{6}.\frac{E_v}{\varepsilon_v}.\frac{d\varepsilon_x}{dr}$$

But in an element of space of energy density "ε_+" in its positive half having an area "A" and length "dr" the total number of positive virtual particles acted upon by "F" is "$n_+.A.dr$", where "n_+" is the number density of positive particles in space. In turn "$n_+ = \varepsilon_+/E_v$". The total force supporting this space element is pressure difference times area and so:-

$$dP.A = F.n_+.A.dr = F.\frac{e_+}{E_v}.A.dr$$

giving:-

$$\frac{dP}{dr} = -\frac{1}{6}.\frac{e_+}{e_v}.\frac{de_x}{dr} \qquad [1.40]$$

This will add to the pressure gradient produced by the exnegmas and then the gradient induced in space itself will act to produce a cumulative effect. Hence the final result can be obtained by substituting either "$(d\varepsilon_-)/dr$" or "$(d\varepsilon_+)/dr$" for $d\varepsilon_x/dr$" in equation[1.40]. The two halves of space act upon both positive and negative particles and when opposites interact the force needs multiplying by "f_v", a number slightly greater than one to allow for the increased perceived volume. Hence the total effect can be summed as follows:-

	FORCE	F	a
-ve half acts on +ve	$-F.f_v$	←	←
+ve half acts on -ve	$+F.f_v$	→	←
-ve half acts on -ve	$-F$	←	→
+ve half acts on +ve	$+F$	→	→

The sum of forces on all particles of space therefore cancel to nothing. This seems strange but if a particle is of either positive or negative mass it will fall in the same direction, as can be deduced by simple logic as well as shown in the last column of the table above. The forces on each are opposite and so the net force on a positive/negative pair would be zero, yet the acceleration in the gravitational field will be identical to that of the separated particles. All particles will tend to fall in the direction of increasing density at the same rate whether they are in positive or negative energy states or composites of both. Hence the tendency is to increase the energy density gradient of both halves of space even though the net energy and pressure is zero. This means that particles of either energy sign tend to fall toward the attractor to cause the density of space to increase progressively. Indeed the system appears unstable so that density gradients will tend to infinity.

Something needs to allow the amplification by space to be very high but prevent it rising beyond some specified limit. This is not hard to find because the particles of space have finite lives. They will accelerate in the direction of the attractor only for a finite time and will then vanish. The build-up in density of space will cause it to go out of equilibrium with the production rate provided by the underlying Grid. So the death rate will exceed production to limit the build-up. It must also not be forgotten that the Grid is not a passive structure. Instead it possesses machine intelligence which can be programmed to sense the local gradient of the exnegmas and adjust production rate so as to achieve a specified amplification factor.

The position has now been reached at which the acceleration, "g" of sub-atomic particles of energy density "ε_p", mass "m_p", energy "E_p" and volume "V_p" can be formulated. With an identical value of "f_v" for all kinds of particle and with "dp/dr" as the pressure gradient of the positive half of space, the net force becomes:-

$$F = -V_p \cdot (f_v - 1) \cdot \frac{dp}{dr}$$

Since g = F/m_p:-

$$g = -\frac{V_p}{m_p} \cdot (f_v - 1) \cdot \frac{dp}{dr}$$

Then noting that "V_p" can be replaced with "E_p/ε_p" and putting "$E_p = m_p \cdot c^2$" and substituting from equation [1.40] with "ε_+" replacing "ε_x":-

$$g^{\cdot} = -\frac{c^2}{6} \cdot \frac{\varepsilon_+}{\varepsilon_p} \cdot \frac{(f_v - 1)}{\varepsilon_v} \cdot \frac{d\varepsilon_+}{dr}$$

[1.41]

Now from equation[1.8] it follows that,
"$d\psi/dr = \phi_m$". This may appear unjustifiable now that
absorption is being disallowed. However, the force must
match the original extended Newtonian physics and the
above is consistent with this need. Alternatively the
case of the single attractor can be considered. Then by
differentiating equation[1.12] "$d\psi/dr = r_o/r^2$". Since
the extended Newtonian physics makes "$F = -E.r_o/r^2$", the
same end result for "ε_1" and therefore "ε_+ is returned.

Also "$d\varepsilon/dr$" is given by equation[1.36] and can
represent the positive half of space. Substituting this
as "$d\varepsilon_+/dr$" in equation[1.41] and with "r_o" from
equation[1.11] and with the attractor mass as
"$m_A = E_A/c^2$" (replacing "E_s" used in [1.11]). Noting that
for a single attractor "$\phi_m = (G/c^4).E_A/r^2$" (from
equation[1.4]), the gravitational acceleration becomes:-

$$g = \frac{e_1}{e_p} . \frac{e^{-3.\psi}}{2} . (f_V - 1) . \frac{e_+}{e_V} . \frac{G.m_A}{r^2}$$

$$\dots\dots\dots\dots[1.42]$$

In the above equation[1.42] "ε_1" is really the same
as "ε_+" because both represent the positive half of
space, the former being referred to the datum position.
This does not really predict "g" because "G", the
gravitational constant, is derived from experiment.
Instead it can be used with "g", also as determined from
experiment, to evaluate the required energy density of
space "ε_+". Hence the above expression is best
rearranged as:-

$$e_+ = -\frac{e_V}{e_+} . \frac{2.e^{3.\psi}}{(f_V - 1)} . \left[\frac{e_p.g.r^2}{G.m_A} \right]$$

$$\dots\dots\dots\dots[1.43]$$

For the weak fields available to observation the term

"EXP(3.ψ)" is unity to a very high order of accuracy.
Even at the surface of a neutron star "ε_1" is only about
1.4 times the value of "ε" taken at infinity.

An inverse square law of force is predicted if "V_p"
is constant. But to match the initial deduction that
"$F = -\phi_m.E$", it is clear that the volume "V_p" of each
elementary particle must vary in proportion to "E_p". It
follows that the energy density of all elementary
particles "ε_p" must be a universal constant. This again
seems a very reasonable and satisfying result.

The volume of an elementary particle will equal
"E_p/ε_p" where "ε_p" is the universal energy density of
elementary particles. If the electron is assumed as an
elementary particle of radius 10^{-18}m and mass
$9.1091.10^{-31}$kg, then $\varepsilon_p = 2.10^{40}$J/m^3. With "g" taken as
9.807 m/s^2, at the surface of the Earth whose mass is
$5.97.10^{24}$ kg and surface radius $6.378.10^6$ m and with
"G" $= 6.673.10^{-11}$ N.m^2.kg^{-2}; then equation[1.43] shows:-

$$\varepsilon_+ = 4.10^{40} . \frac{\varepsilon_v}{\varepsilon_+} . \frac{1}{(f_v - 1)} \qquad \frac{J}{m^3}$$

Now "$\varepsilon_v/\varepsilon_+$" must be greater than about 4 otherwise the
virtual particles of space would jam into a solid mass.
A better guess would be about 100. This yields a value
of "ε_+" still considerably smaller than the value of
10^{45}J/m^3 given by Starobinskii and Zel'dovich(215). Their
value is presumably based on the mediator requirements
for the electric force, since a figure very close to
this is predicted in Chapter T.S.3 from consideration
of the electric force alone.

Then in order to provide a match it is necessary to
make "$(f_v - 1)$" equal to 0.01. This means that when
opposite energies interact they see a volume about 1.01
times larger than for the case of like-energy
interactions. Again this does not seem an unreasonable
figure. Hence provided the value quoted above for "ε_+"
can be substantiated by values obtained in the next
chapter for the electric force, the two long range
forces will have been related to one another to within

striking distance.

A somewhat unexpected feature is shown, however. The energy density of sub-atomic particles has to be very low in comparison with the virtual mediating particles of space. According to this analysis the latter need a density about 5.10^6 times that of sub-atomic particles! So the latter must be complex structures made of energy greatly "fluffed out" as compared with the ultimate density. The matter is dealt with further in Chapter T.S.3 where the energy density of space needed to explain the electromagnetic force is derived. There it is shown that the electric force can also be explained by a buoyancy model.

It does not seem unreasonable, in view of the favourable result given by this analysis, to suggest that much progress toward relating the strengths of the gravitational and electric forces has already been made.

1.6 CONCLUSIONS.

A theory of quantum gravitation has been briefly described which satisfies all the experimental checks which have been made to establish Einstein's theory of general relativity except for one which has not yet been fully addressed. This is the quantification of gravitational waves. The latter are predicted by the new theory but so far the energy radiated by gravitational wave production has not yet been formalised.

Only basic theory has been concentrated upon but any mathematician would be able to derive all final expressions where detailed derivations have been omitted. These derivations are too lengthy for inclusion but will be given in full in the technical book (212)

The new theory yields the same equations for the gravitational red shift and the same doubling of bending of a beam of light as Einstein's theory. It gives a prediction for the Shapiro time delay so close to

Einstein's prediction that it is not possible to discriminate between the two by observation. It gives an apparently quite different equation for the precession of the perihelion of planets, yet, surprisingly, almost exactly the same predictions are made for any conceivable orbit in the weak fields available to observation. Sufficient data has been supplied in FIG.13* for the reader to be able to make independent checks. The equations give the answers in radians and so a conversion to arcseconds is required.

Finally the new theory explains the null experiments of Eötvös and Dicke just as well as does "equivalence".

The new approach, however, achieves a goal which relativity is unable to reach. This is the prediction of the ratio of the electric to the gravitational force to within striking distance. It shows why the ratio is as large as it is.

Furthermore there are no internal contradictions. The black hole does not exist in the new theory because the speed of light does not fall to zero at a finite radius about a massive object as it does in relativity. Instead if it were possible to concentrate a massive object within a singularity, then the speed of light would only fall to zero at zero radius. Hence no longer does any difficulty exist. The problems associated with the black hole, such as the breakdown of logic do not arise.

There is no inconsistency with the idea of negative energy states. With these included the problem of a huge but non-existent cosmological constant is absent. In established theory resort is still being made to other universes existing in higher dimensions which connect to ours via so-called "Wormholes" in space. Yet at the same time the "negative coupling" used to explain attractive forces was shown in Chapter 5 to also contain a contradiction.

It can be concluded that the established approach has to find a way to accept negative energy states if it is to advance to the stage at which it can be regarded as at all plausible. At the present time such solutions

* See Page 280

are prohibited because they cause embarrassment in other parts of quantum theory.

In the new theory light propagates relative to local space. The structure described for the electromagnetic wave had to be based on this assumption. It is necessary to point out that existing theories for this wave are all purely abstract and make no attempt whatever to explain how a neutral photon can transmit electromagnetic effects. In special relativity light propagates independently of any medium. This assumption is therefore incompatible with the existence of the quantum vacuum. Hence an internal contradiction now exists in the present solution. It is bound to exist in any quantum-based theory owing to the way the quantum vacuum is defined. The new theory started out by assuming special relativity to be valid, so that the initial equations[1.1] and [1.2] were taken to hold.

Clearly the inconsistency just pointed out will invalidate the entire derivation, unless the two equations assumed can be derived by an alternative which is based on the assumptions given in section 1.2 for the extended Newtonian physics. It will be necessary to provide a modified theory of special relativity which nevertheless meets all the experimental constraints and yet is totally free from internal contradiction.

This task is attempted in the next chapter. It will be seen to succeed.

CHAPTER T.S.2

MODIFIED SPECIAL RELATIVITY
AND THE QUANTUM VACUUM

2.1 INTRODUCTION - THE PROBLEM

When space is treated as an ideal compressible superfluid, a promising solution appears for difficulties arising when special relativity is integrated with quantum gravitation.

Einstein based his theory of special relativity on the postulate that light propagated at the same speed "c" in vacuum, according to all observers placed in unaccelerated frames of reference. Such frames of reference were called "inertial systems" and it needs to be emphasised that accelerations were not considered. This postulate was consistent with the observations of Michelson and Morley.

They tried to detect differences in the speed of light in different directions and found them to be zero. This null result seemed to eliminate the possibility of the Earth moving through an "ether". At the beginning of the century, the ether was regarded as a rigid all-pervading substance through which matter would need to travel and in which light would propagate. The possibility that this hypothetical medium could be dragged along with the Earth had been ruled out by the astronomical observation of "stellar aberration". Starlight arrived with a slight inclination which made all the stars appear to move with a common orbital motion. This was consistent with the Earth moving through the ether with a speed equal to that of its

motion round the Sun. This incompatibility was taken to
mean that the concept of an ether was flawed. A third
option, that light would propagate at a speed "c"
relative to the emitting object, could not be
entertained because it would be inconsistent with
observations on eclipsing binary stars. Because one
member is moving towards us whilst the other recedes,
light would take longer to reach us from the receding
member and observations show this to be impossible.
Only   Einstein's   concept   of   light   propagating
independently of any medium appeared valid.

In Chapter T.S.1 it was shown that special
relativity could be applied at all levels in a
gravitational field. However, there has always been a
difficulty and it has been left to this chapter to find
a solution. Indeed this late stage parallels the
historical development of the theory of quantum
gravitation being presented, because the problem seemed
more intractable than any of the others and took in fact
nearly four years to resolve.

Light moved independently of any medium in
Einstein's theory and this feature leads to a logical
inconsistency when attempts are made to derive any kind
of quantum theory of gravitation. As shown in the
previous chapter, light needs to propagate relative to
local space, since photons could not possibly penetrate
the quantum vacuum in which virtual electric charge is
present. Indeed it will soon be shown that there are
several internal inconsistencies within special
relativity itself. These alone give sufficient cause
for concern to make it necessary to search for another
solution.

This will have to accept the existence of space
described as the "quantum vacuum", needed to account for
the three remaining forces of nature. Such space is
considered to be a seething mass of virtual particles,
each having only a fleeting existence. Furthermore, it
was shown in Chapter T.S.1 that in order to visualise
true space and quantify its compressibility it was
expedient to consider a simple model for the
electromagnetic wave. This postulated that photons

jumped repeatedly from one virtual electron to another.

The space described by quantum theory would not be rigid like the ether; instead it would behave more like a fluid, since the virtual particles do not cohere in the manner of atoms in a solid. Mathematical treatments of relativity and gravitation such as that given by Møller(210) never discuss fluid mechanics. It appears that the possible relevance of this discipline has been totally neglected. Yet according to Møller in 1851 Fizeau measured the velocity of light in running water, followed by confirmation in 1886 by Michelson and Morley. These experiments carried out by sending one beam upstream and one downstream before joining them to produce interference, showed light to propagate with a bias in the direction of flow. The proposals now introduced are consistent with such a result, replacing one fluid -water by another -superfluid space.

In Chapter T.S.1 space was shown to be compressible. This chapter will study the effect of regarding space as a compressible superfluid. It will be shown that it is then possible to find a solution in accord with all observation. Briefly, a local bubble of space will be shown dragged along with the Earth but this moves through a larger space-bubble locked onto the Sun. This "Sun space" flows over and around the Earth bubble. The change from the concept of a rigid ether to a fluid space makes all the difference, as will be shown.

The concept is compatible with gravitation resulting from buoyancy-type forces. Here space behaved as a superfluid on a sub-atomic scale. A modified form of special relativity will extend the principle to the macroscopic scale.

First the equations[1.1] and [1.2], normally derived from special relativity by imagining bodies moving at constant speeds comparable with light, will be obtained by considering acceleration in a manner consistent with the assumptions of the extended Newtonian physics. Then the assumptions of special relativity will be reviewed and compared with a modified set. Finally some new experimental checks will be proposed.

2.2 "RELATIVISTIC" EQUATIONS FROM ACCELERATION

Special relativity imagines objects in uniform motion and zero gravitational field and no details of atomic or nuclear structure are considered. Fortunately, by studying the effects of acceleration on matter, as required to produce the relativistic speed differences considered by Einstein and taking account of atomic structure, an alternative approach can be adopted which yields the same basic equations. This can be achieved by considering the example of a mass driver capable of accelerating objects to relativistic speeds.

The logic can be summarised as follows:-

Light is deflected in a gravitational field, as shown by stars near the limb of the Sun during solar eclipses. Now the photons of light behave as if they consist of kinetic energy alone. Consequently energy behaves gravitationally like mass and so the two must be equivalent. It may be objected that this represents a circular argument. It may be contended that until the required expressions are deduced, there is no way of knowing that light behaves like kinetic energy.

Another starting-point can be used instead which avoids this difficulty. When atomic nuclei experience radioactive decay they split into two smaller nuclei. The products are always emitted with high speed and so carry away high kinetic energy. It is always found, that the sum of the masses of the products is always smaller than that of the original nucleus. It can be inferred, therefore, that the mass loss must in some way be proportional to differences in binding energies of the nuclei. Furthermore these mass differences must be related to the initial kinetic energies of the products. This deduction, it should be noted, owes nothing to relativity.

The null experiments of Eötvös and Dicke provide further support. These showed that gravitational force is always proportional to the inertia of a body, regardless of any difference in binding energies. Objects of different materials, known to have different nuclear binding energies, were placed at the ends of a

beam suspended horizontally from a single fibre. This
formed a torsion balance of extreme sensitivity. Placed
at mid-latitudes on the surface of the Earth the objects
would feel a horizontal component of centrifugal force
caused by the Earth's rotation. The more sensitive test
of Dicke used the gravitational pull of the Sun, which
reversed its direction every half-day, to affect the
balance of inertia force caused by Earth rotation. The
two masses were arranged to lie in the East-West
direction. If gravitation did not respond to binding
energy, then a torsional oscillatory deflection with a
24-hour period would have been detected. Null results
were returned to an accuracy of one part in
100,000,000,000 according to Will(12). But he also
states that these checks, which were carried out to
prove the "equivalence" of gravitation and acceleration,
could only be explained by curved geometry
interpretations of gravitation.

This contention is disputed here because the
interpretation presently considered fits the facts
equally well and depends only on Euclidean geometry.
The result is interpreted here to mean that the
gravitational force is affected by any binding energies
in exactly the same proportion as inertia is affected.
Since inertia is a property of mass and gravitation has
already been shown proportional to the product of total
energies, it follows that mass differences and energy
must in some way be equivalent to one another.

It is therefore permissible to say that a constant
of proportionality "B" exists so that a given mass
"m" can equally be represented by an energy "E" such
that "m = B.E".

Now Newton states that the force produced on a body is
equal to its rate of change of momentum, or:-

$$F = \frac{d(m.v)}{dt} \qquad [2.1]$$

It follows that an equivalent is:-

$$F = B. \frac{d(E.v)}{dt} \qquad\qquad [2.2]$$

Expanding [2.2] by differentiation by parts we obtain:-

$$F = B.\left(E. \frac{dv}{dt} + v. \frac{dE}{dt}\right) \qquad and \quad v = \frac{dx}{dt}$$

So nothing is changed if both sides are multiplied by

$$\frac{v}{\left(\frac{dx}{dt}\right)}$$

i.e.:-

$$F = B.\left(E. v. \frac{dv}{dx} + v^2. \frac{dE}{dx}\right)$$

Now the work done by force "F" in an element of distance "dx" is "F.dx" and is equal to the gain in kinetic energy "dE" of the object. This element is added to the energy "E" already existing, so "E" must increase as the particle accelerates. Hence, multiplying both sides by "dx" we have:-

$$F.dx = dE = B.(E.v.dv + v^2.dE)$$

And if "E_o" is the "rest energy" of the object, which is the value at "v = 0", this can be re-arranged as:-

$$\int_{E_o}^{E} \frac{dE}{E} = B.\int_{o}^{v} \frac{v.dv}{1 - B.v^2}$$

Now this is a standard form and can be integrated by putting "$z = 1 - B.v^2$", so that "$dz = -B.2.v.dv$". Integration yields:-

$$\ln\left(\frac{E}{E_o}\right) = -\frac{1}{2} \cdot \ln(1 - B \cdot v^2)$$

Or

$$\frac{E}{E_o} = (1 - B \cdot v^2)^{-\frac{1}{2}} \qquad\qquad [2.3]$$

As "v" is increased so "E" increases according to equation [2.3] until "B.v^2 = 1". At this point "E" becomes infinite and so no further increase of "v" can arise no matter how large the accelerating force "F" or how long it is allowed to act. An electron accelerating in this way is observed to reach a speed asymptotic to the speed of light "c". Hence "B = $1/c^2$". Also particles like photons can only travel at speed "c". The only condition which satisfies [2.3] is then "E_o = 0". Hence photons can only have a rest energy of zero and they can only consist of kinetic energy.

Also substituting B = $1/c^2$ in [2.3] it follows that:-

$$E = \frac{E_o}{\sqrt{1 - \left(\frac{v}{c}\right)^2}} \qquad\qquad [2.4]$$

And the same substitution in [2.2] equated with [2.1] yields:-

$$F = \frac{d(m \cdot v)}{dt} = \frac{1}{c^2} \cdot \frac{d(E \cdot v)}{dt}$$

Integrating it follows that:-

$$m \cdot v = \frac{1}{c^2} \cdot E \cdot v$$

Which simplifies to:-

$$E = m \cdot c^2 \qquad\qquad [2.5]$$

Equations [2.4] and [2.5] are usually derived from the

special relativity for which Einstein is famous. He imagined objects moving at steady speeds. No accelerations were considered and so the need to equate force to rate of change of linear momentum did not arise. But the new derivation owes nothing to relativity. It began from Newtonian physics and **depended** on acceleration. The starting-point of the derivation of quantum gravitation given in the previous chapter need not have postulated these equations set out as [1.1] and [1.2] respectively. A start could have been made with this derivation and then no mention of special relativity need have been made.

The increased inertia of accelerated sub-atomic particles caused by increased total energy also predicts time dilation. This aspect will be considered next.

2.3 "RELATIVISTIC" TIME DILATION

The successful derivation of equations [2.4] and [2.5] encourage further use of the mass driver example, this time concerning the dynamics of atoms. Any kind of atom is to be assumed accelerated. The atom is described by the Schrödinger wave equation but the Bohr radius "a_o", strictly applicable only to hydrogen, corresponds with the position of maximum probability for distance of the electron from the nucleus. It is therefore permissible to use "a_o" as a reference radius to define the size of an atom. At this radius the electron will have a tangential speed "u_1" when the atom to which it belongs is at rest. This will change to "u" when the atom is accelerated to a linear speed "v". The Schrödinger model appears to describe an atom consisting of a virtual electron associated with permanent energy and momentum. It sparks in and out of existence at a very high frequency but, arising in a new place after each extinction, conserves these properties. It is therefore related to the simpler Bohr model, which is easier to adopt for the present analysis.

In the Bohr model a permanent electron makes a circular orbit of radius "a_o". This radius will be nearly the same for all atoms when only the outermost

electron is considered, since the binding forces of all
but one proton of the nucleus are cancelled by the other
electrons.   The effect of accelerating the electron
bound to the atom will be investigated.   No angular
momentum is able to be imparted to the electron by
linear acceleration of the whole atom and so it is
possible to write:-

$$m.u.r \;=\; m_1.u_1.a_o \qquad\qquad [2.6]$$

Now if the atom is accelerated to a linear speed "v" and
if for simplicity the axis of the electron's orbit is
made to coincide with the direction of "v", then the
absolute speed "w" of the electron will be given by:-

$$w^2 = u^2 + v^2 \qquad\qquad\qquad [2.7]$$

For any other orientation of the electron orbit "w" will
be the average velocity, because if linear acceleration
cannot change angular momentum, neither can the
electron's orbital energy be altered. Hence orientation
can have no effect. Also the values of "m" and "m_1" can
be related to the rest mass "m_o" of the electron by
combining equations [2.4] & [2.5] so that equation [2.6]
becomes:-

$$\frac{u}{c} \;=\; \frac{u_1}{c}.\frac{a_o}{r}.\frac{\sqrt{1 - \left(\frac{v}{c}\right)^2 - \left(\frac{u}{c}\right)^2}}{\sqrt{1 - \left(\frac{u_1}{c}\right)^2}}$$

Which can be re-arranged to yield:-

$$\frac{u}{u_1} \;=\; \frac{\sqrt{1 - \left(\frac{v}{c}\right)^2}}{\sqrt{\left(\frac{r}{a_o}\right)^2.\left[1 - \left(\frac{u_1}{c}\right)^2\right] + \left(\frac{u_1}{c}\right)^2}} \qquad [2.8]$$

If "r/a_o" remains unity, then in the above the terms
"u_1/c" eliminate and the orbital speed will reduce as
"v/c" is increased and in exactly the same ratio as the

time dilation factor obtained by special relativity, i.e:-

$$\frac{t}{t_1} = \frac{u_1}{u} = \frac{1}{\sqrt{1 - \left(\frac{v}{c}\right)^2}} \qquad [2.9]$$

It is reasonable to treat time dilation as the inverse of orbital frequency, which is implicit in the derivation just presented, since this factor will determine all rates of atomic clocks. The mass driver can be studied further to find how "r" changes with "v".

2.4   CHANGE OF DIMENSIONS WITH ACCELERATION

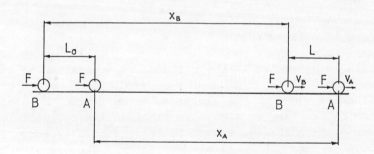

FIG. 16 <u>TWO ATOMS EQUALLY ACCELERATED.</u>

Equation[2.4] can be re-arranged as :-

$$\left(\frac{v}{c}\right)^2 = 1 - \left(\frac{E_o}{E}\right)^2$$

Then with "$\delta E = F.x$"   and   "$E = E_o + \delta E$" the above becomes:-

$$\frac{v}{c} = \sqrt{1 - \frac{1}{\left(1 + F.\frac{x}{E_o}\right)^2}}$$ [2.10]

Two identical atoms "A" and "B" are to be considered, separated initially by a distance "L_o" in the "x" direction. They are both acted upon by an equal force "F" at the same instant "t = 0" and since "E_o", the rest energy of each atom, is the same the parameter "$F.x/E_o$" will be equal for both. It follows that after the lapse of an equal time interval "t" for both the velocities reached, "v_A" and "v_B" respectively given by equation[2.10], will be equal. The average velocities will also be equal and consequently the values of "x_A" and "x_B" will be equal. Now from FIGURE 16 it is clear that:-

$$x_A + L_o = x_B + L$$

Hence it follows that: $L = L_o$

There is no change in the distance of separation of the two atoms. The atoms could be replaced by electrons without affecting the argument. The force "F" could then be provided by a uniform electric field capable of accelerating the electrons to relativistic speeds. It would indeed be possible to check the result by experiment, using a laser to trigger emission from a pair of cathodes separated by a distance corresponding with "L_o". However, the logic is so simple that such a check is hardly necessary.

But if "A" and "B" are atoms, they could be connected by a string of other atoms bound by their chemical bonds and able to carry force along the chain from the leading end point, marked by atom "A", to the trailing one, marked by atom "B". Clearly if each atom is acted upon by its own individual accelerating force, then it will be unaffected by the others and "L" will remain constant. It is clear that reference could now be made to a object of macroscopic size. This proves that the length measured in the direction of motion cannot be affected by acceleration to relativistic

speeds unless some other force acted which caused atoms
to contract in size.

2.4.1   ACCELERATION OF A CUBE

Next some way needs to be found for finding how
lateral dimensions will vary with acceleration to
relativistic speeds. This can be achieved most easily by
considering the acceleration of a cube. This need not
be of stiff material. In the first case to be
considered it can be represented by eight isolated
spheres one arranged to represent each corner of the
cube and all of identical mass. FIG.17 illustrates the
logic.

Use is again made of a hypothetical mass driver, but
this time it is made in two sections. The first
produces an acceleration in the "x" direction by adding
kinetic energy "δE_x" to rest energy "E_o". Then kinetic
energy "δE_y" is added in a perpendicular direction such
that its magnitude is equal to that of "δE_x". Each
corner-sphere is assumed to be acted upon by forces
which are exactly equal to forces acting on any of the
others, so all are subjected to identical accelerations.
But in the present case the two energy components have
equal magnitude. Hence it follows from symmetry that
the corresponding velocity components will be equal,
i.e. "$q_x = q_y$". This means that the final direction of
motion of all spheres is at 45 degrees to the "X" axis
and corresponds with the direction of a diagonal drawn
across the "x-y" face of the cube.

Now the final vector sum velocity "w" could have
been produced in a single step by a linear acceleration
directed at 45 degrees to the "x" axis. Hence the
diagonal of the cube-face will correspond with "L",
which has already been shown to remain constant during
acceleration. Since also "L_x" has already been proved
constant, it follows that the transverse dimension "L_y"
will not change either. The argument can be repeated in
the third dimension.

Hence the cube of isolated spheres will remain a
cube of constant size and shape when accelerated up to
the speed of light. If it were a solid cube made from

atoms, and if each atom contracted with increase of
speed, then the cube would contract as a whole but would
remain constant in shape.

This conclusion is at variance with the predictions
of special relativity. As a consequence of its initial
assumptions this theory yields a contraction in length
so that:-

$$L = L_o \cdot \sqrt{1 - \left(\frac{v}{c}\right)^2} \qquad [2.11]$$

The "Lorentz contraction", "$L_o - L$", is given by
incorporation with equation[2.11]. This applies only to
dimensions in the direction of motion, the transverse
dimension in special relativity does not change. Hence
in this case objects are distorted. If the cube example
is repeated using the assumption that contraction arises
in only one direction then an internal inconsistency
immediately becomes apparent.

Motion of a huge cube is to be assumed at speeds
close to that of light and viewed from a great distance.
FIG.17 is relevant since no velocity component arises in
the direction of line of sight. Acceleration of the
cube in the "X" direction takes place first and gives a
shortening in "L_x" but not "L_y". Then adding an equal
energy by acceleration in the "Y" direction gives a
shortening in "L_y" with only minor change in "L_x". There
will be a slight recovery in "L_x" because the velocity in
the "X" direction will reduce somewhat when energy in
the "Y" direction is supplied. This can be ignored for
the present argument because some contracion in the "X"
direction will remain. The final result is a cube of
smaller dimensions.

Carried out as a single process by accelerating
along the diagonal, however, all the shortening takes
place along the diagonal and the cube distorts to a
rhomboid shape. The same end result needs to arise
whatever series of processes are employed and so
something must be wrong.

It should be noted that in the derivation of
equations [2.1] to [2.8], no reference was made to the

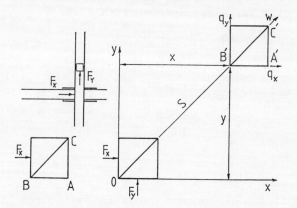

FIG. 17   ACCELERATION OF A CUBE

way light would behave.   They are therefore equally
applicable to any theory, inclusive of special
relativity.   Clearly, therefore, a serious internal
logical inconsistency has been demonstrated to exist in
the theory of special relativity as a result of
Einstein's initial postulate.   Hence his assumptions
must be false!

It is in fact impossible to deduce by logic that
cubes of stiff material will remain of constant size
when accelerated, even though their shapes will not
distort.   This is because atoms need now to be imagined
filling the spaces between the eight separate spheres
previously considered.   The atoms might contract and
exert a small force, causing a large object to contract.

If it is **assumed** that size will not change then
"r/a_o" remains unity in equation[2.8] so that the
observed time dilation given by equation[2.9] will
apply.   Then if the electric force is assumed to control
the value of "r" it is possible to determine the way
electric charge must vary with translational speed.   The
value of u_1/c is very small for electrons in the outer

energy shells of atoms and for this case where "e" represents electronic charge the electric binding force "F" for a circular Bohr type orbit is given by:-

$$F = k.\left(\frac{e}{r}\right)^2 = m.\frac{u^2}{r} \qquad [2.12]$$

It is then simple to show that for the orbiting electron, when "r" is fixed and equal to "a_o":-

$$\frac{u}{u_1} = \frac{m_1}{m} = \left(\frac{e}{e_1}\right)^2 = \sqrt{1 - \left(\frac{v}{c}\right)^2}$$

$$\ldots\ldots[2.13]$$

Hence the electronic charge needs to fall slightly as speed increases. This is unlikely to be the case otherwise it must surely have been discovered by now during experiments, although this is well worth checking.

At any value of "v/c" during acceleration of a charged particle in an electric field, the next element of volt-drop "δE_v" will need to be increased to "$\delta E_v.[1 - (v/c)^2]^{-1/4}$". It can then be shown by integration that to accelerate from rest to a given "v/c" requires a volt drop increase ratio "$\delta Ev'/\delta Ev$" of:-

$$\frac{\delta E_v'}{\delta E_v} = \frac{2}{3}.\frac{\left[1 - \left(\frac{v}{c}\right)^2\right]^{-3/4} - 1}{\left[1 - \left(\frac{v}{c}\right)^2\right]^{-1/2} - 1} \qquad \ldots[2.14]$$

This shows that to accelerate an electron to "v/c" = 0.9 the potential drop needs to be increased 1.275 times the 661,330 volts needed with constant charge. When "v/c" = 0.99, the potential needs to be increased in the ratio 1.957 so that instead of requiring $3.11.10^6$ volts, $6.09.10^6$ would be needed.

Alternatively with constant charge the atom will shrink and then the observed time dilation will not arise. Since constant charge seems the most probable we are therefore still not yet "out of the wood". In this case a paradox has arisen.

The solution can be found from the main text. The electron does not excecute Bohr type orbits and does not need the electric force to structure the atom! It uses an abstract wave-number control system to specify the point of appearance of any electron of its sequence. The force needed to curl the wave function into a ball also only needs to be an abstraction. It does not need to relate to the real electric force at all. The reason there is a match, as indicated by the matching size of the Bohr orbit, is therefore merely fortuitous. Or more probably a designer's choice, now that this theory shows a Creator to have been involved. The real and abstract forces simply drift out of step with each other as speeds increase. The electric force is needed for other purposes such as causing atoms to stick together and for the propagation of light.

The electrical stiffness of the atom will also be governed by the abstract force and so will vary as "$[1 - (v/c)^2]^{1/2}$" and is the same ratio as the inverse of the mass ratio. Since vibration frequency is proportional to "$(stiffness/mass)^{1/2}$" the correct time dilation effect as given by equation[2.9] is predicted and neither atoms or larger structures will change in size as they accelerate. This is despite the assumption that the true electric charge does not change.

Since the relativistic time dilation has already been confirmed by experimental observation it follows that the size of atoms is fixed. No check has been made to test the Lorentz contraction, however, and as already shown this cannot arise.

Since unstable cosmic rays would decay before reaching the Earth's surface unless the same time dilation factor applied, it can also be deduced that composite particles like muons and nucleons will not change in size when accelerated nearly to the speed of light. In such cases it is the rotating quarks, bound by the strong force, which need to be considered instead of electrons.

It is now necessary, however, to find an assumption

for the propagation of light which will be consistent
with these predictions, special relativity having been
found to be inconsistent. Already we know what this
needs to be. Light needs to propagate at a speed "c"
which is fixed relative to local space in zero gravity,
measured in the units of time and distance at rest with
respect to such space. A simple set of checks for
consistency will now be attempted.

2.4.2 POSITION DIAGRAMS FOR CHECKING RELATIVITY

A contraption or space-ship of great length in the
"X" direction is to be imagined accelerated, in the "X"
direction only, by the mass driver until a steady speed
"v" is reached. It has a flash tube centrally located,
which is able to send a short pulse of light both
forwards and backwards to meet mirrors angled at 45
degrees, so that the reflected beams emerge
perpendicular to the direction of motion. The object is
to measure the length of the space-ship as it flies
past, by capturing images on film using a camera fixed
relative to local space.

In order to obtain an accurate result both light
pulses need to arrive simultaneously at the camera.
This is necessary because if a time delay were to be
allowed, then one mirror would have moved on in the
interval, leading to an error in length measurement. It
is possible to analyze the result using a frame of
reference having any speed relative to local space. Two
cases will be investigated, one at rest, the other with
the observer moving with the space-ship.

Consider the rest frame first, as shown in FIG.18.
A position diagram is drawn in which distance is plotted
horizontally with time moving vertically upwards.

2.4.3 FRAME AT REST IN LOCAL SPACE

The speeds of propagation both forward and backward
are equal to "c". The forward pulse needs to travel a
distance increased by amount "$v.\delta t_A$", due to the speed of
the space-ship "v" and the travel time of the pulse
"δt_A". On the other hand the rearward pulse travel time
is shortened and it is therefore necessary to delay the

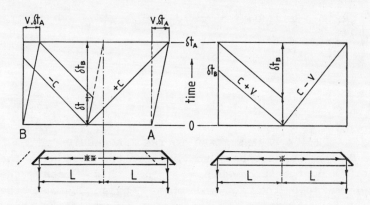

FIG. 18 VIEWING FRAME AT REST.   FIG. 19 OBSERVER MOVING

MOVING CONTRAPTION. (SPACE SHIP)

rearward flash by time "δt", so that both arrive
simultaneously at the mirrors. Hence from the time
pulse "A" begins, the distance of travel of the rearward
pulse is shortened by amount "$v.\delta t_A$". Hence it can be
written:-

$$L - v.\delta t_A + v.\delta t = c.\delta t_B \quad \& \quad L + v.\delta t_A = c.\delta t_A$$

$$\& \quad \delta t = \delta t_A - \delta t_B$$

$$So \quad L - v.\delta t_B = c.\delta t_B$$

Hence the two cases become:-

$$\delta t_B = \frac{L}{c + v} \quad \& \quad \delta t_A = \frac{L}{c - v} \qquad [2.15]$$

The time delay "δt" from A to B flash is therefore:-

$$\delta t = \delta t_A - \delta t_B = \frac{L}{c - v} - \frac{L}{c + v} \quad or:-$$

$$\delta t = \frac{2.v.L}{1 - \left(\frac{v}{c}\right)^2} \qquad [2.16]$$

2.4.4 FRAME FIXED TO THE SPACE-SHIP

Now it is necessary to consider the case with the observer on the space-ship. In this case the observer sees the forward pulse moving against the motion "-v" of space, so the apparent propagation speed is "(c - v)". The rearward pulse moves with this motion at relative speed "(c + v)". Hence it follows that:-

$$\delta t_B = \frac{L}{c + v} \quad \& \quad \delta t_A = \frac{L}{c - v} \qquad ...[2.17]$$

which is exactly the same as the previous case, (equation [2.15]) leading to the same delay time "δt". It needs to be noted that the unit of time is that of the previous case. The moving observer works in dilated time owing to the increased inertia caused by energy addition. Hence both "c" and "v" appear to have relative speeds increased in the ratio "$[1 - (v/c)^2]^{-1/2}$". The speed seems greater than light in the rearward direction.

It also follows that for either frame of reference the length of the space-ship "2.L" will measure the same in motion as at rest when the pulses arrive simultaneously at the camera. This result is exactly in accord with the conclusion arrived at by considering acceleration.

2.5 THE EINSTEINIAN CASE

The case with the frame at rest looks exactly like FIG.18 except that "L" has shrunk to "L'" due to the Lorentz contraction. The delay time needed will also be reduced in the same ratio. But for the case of the moving observer (mounted on the space-ship) the propagation speed appears to be "c" in both directions and so the time delay for the two flashes is zero. This is an example of "non-simultaneity".

Now two events which happen instantaneously, as would be observed from a background "Grid" with infinite speed of information transport, can be expected to appear non-instantaneous to an observer not equidistant from each because of the difference in time taken for light to travel. But in the example, the stationary observer is equidistant from the two mirrors at the instant of observation described here. It is very difficult to accept that under these circumstances two events equidistant from both moving and stationary observers could appear instantaneous to one and not the other. Nor does it seem reasonable that a time delay "δt" can be required for firing the rearward pulse "B" with respect to the forward one "A" when observed from the stationary frame, whilst for the observer moving with the space-ship there should be zero time delay between the pulses.

It is accepted in special relativity that this will happen, but can it really be true?

It has already been conclusively proved that objects moving at relativistic speeds will not be contracted in the direction of motion. The basic assumption of special relativity, however, leads directly to the prediction of the Lorentz contraction; it follows that this assumption is invalidated. This inference is strengthened by the demonstration given at the start of this chapter, that the major equations of special relativity can be readily derived by studying acceleration, without using Einstein's assumptions and without reference to light-beams.

However, in special relativity the result of the Michelson - Morley experiment does not conflict with the observation of stellar aberration. It is necessary, therefore, to find a way of explaining the same effects using the modified theory under investigation.

2.6 MODIFIED SPECIAL RELATIVITY

It does not in fact appear to be possible to formulate a modified theory, equivalent to special relativity, in isolation from gravitational considerations. This is because the explanation to be offered for stellar aberration is integrated with gravitational effects. The new concept of gravitation will first be summarised.

Mediators are emitted from around all elementary particles in accord with accepted quantum theory. Mediators can, however, only account directly for a minute fraction of the observed gravitational effect. This is because they derive from what can be called the "exnegmas", shown in Chapter T.S.1 to be limited to negative energy states; their total being equal and opposite matter with which they are associated. The mediators of gravitation may only be some fraction of the total exnegmas. They can therefore only provide a primary cause for gravitation by acting upon the remaining virtual particles of space. The latter is thereby compressed slightly. So huge is the gross energy density of space, however, Starobinskii and Zel'dovich(215) giving a value of 10^{45} J/m3, that this has an amplifying effect which dwarfs the direct effect of the mediators alone. It is the pressure gradient of space, caused by compression, which acts across the volume of each elementary particle of matter in the manner of a buoyancy force, which produces gravitation. Indeed all the effects which are attributed to the curvature of space-time in general relativity are exactly parallelled, for the weak fields available to observation, by the effects of compressibility.

The observation of stellar aberration is also predictable, however, when space is considered as an ideal fluid, that is one having zero viscosity like liquid helium. And this *requires* light to be propagated relative to local space.

A text such as that of Vallentine(216) shows how streamlines can be plotted by use of the idealised "source and sink" method. Also, standard flow patterns can be superimposed. For the case of mediators diffusing through space the same method gives a flow net in which surfaces of "equipotential" represent those of uniform pressure and surfaces perpendicular to these represent net directions of flow governed by resistance of the medium.

Applied to space, the pressure drops are negative because the mediators have an attractive, not a repulsive, effect and so space is induced toward ponderous masses. Spherical symmetry prevents motion and so space is pulled toward matter. It is compressed within and around matter. This describes the field of the Sun. The Earth will also have its field, locked around it by mediators. Mediators from the Sun will stream radially outward, supplemented locally from almost the point source of the Earth. Some from the latter will move initially toward the Sun until they reach a "stagnation point" and are turned around, so that all eventually stream out almost radially with mediators from the Sun. There will be a slight deflection of about 10^{-4} radians owing to the orbital motion of the Earth. At the stagnation point the mediator flux contributions from Sun (S) and Earth (E) are equal. Indeed, a surface of "equal mediator influence" can be defined surrounding the Earth and given by:-

$$\frac{m_S}{R_S^{\,2}} = \frac{m_E}{R_E^{\,2}} \qquad\qquad [2.18]$$

This will be almost a sphere of radius "R_E" = 260,000 kilometres, struck from the centre of the Earth.

Within this sphere, the Earth's "space bubble", mediators from Earth will be controlling. So this space will appear at rest, so satisfying the Michelson-Morley null experimental result.

The mediator diffusion pattern is illustrated in FIG.20. The case shown is plotted for two-dimensional flow and so is not strictly accurate. It is good enough for purposes of illustration and the three dimensional case is not very much more difficult to calculate. The mediator flux from Earth is shown by radial lines, which

are numbered.   The flux from the Sun is shown as
parallel lines because the Sun is so far away.   These
are also numbered.   Each streamline is joined to points
in which the sum of the numbers from each set of lines
is constant.   Running in a direction perpendicular to
the streamlines are lines of equipotential, although not
shown in the illustration.   For the diffusing flow
pattern which the model represents in this case the
equipotentials are real surfaces of uniform pressure and
the differentials between them provide the driving force
for mediators.   They flow through gross space, which
acts rather like a porous substance.   They "diffuse"
through space.

The space around the bubble will be controlled by
the Sun, immersed in its own space bubble of huge size
and encompassing the entire solar system.   The Earth is
in motion through Sun-space.   Clearly there are two flow
systems at work having totally separate patterns.   The
mediator flow pattern has been described as a net
radially outward diffusion centred on the Sun.   But the
two major halves of space cannot have a radial flow;
part is locked around the Earth, that which is confined
within the local space bubble, and the rest is locked
around the Sun.

Owing to the orbital motion of the Earth, it follows
that a surface of shear must exist and this will
approximate to equation[2.18], since this gives the
condition of equal mediator control from Sun and Earth.
It is therefore the surface of greatest weakness.   The
gross space of the Sun, seen from Earth, will therefore
flow over the boundary of Earth's bubble, just as if it
were an ideal solid object.   Or just as accurately, as
if over the dividing surface of a "doublet".   This is
the pattern obtained by adding a source to a sink at the
same place and superimposing a uniform flow field.   The
point of importance to note is that the external field
does not penetrate into the bounding surface or create
a drift flow inside.   In addition, for a superfluid a
shear zone of negligible thickness can exist without any
drag force being transmitted, because the fluid has zero
viscosity.   In this case the boundary surface can

contain fluid space
having no relative
motion with respect to
the central binding
mass. So there will be
no motion inside the
Earth bubble except for
a low-speed rotation,
somewhat like a free
vortex with tangential
velocity decreasing
with distance, caused
by the spin of the
Earth about its polar
axis.

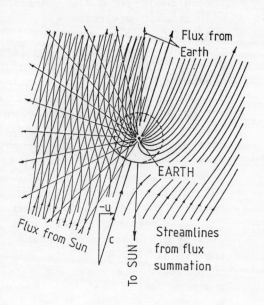

No turbulent
boundary layer will
form at the surface of
the bubble as would
arise with viscous
fluids. It can be
inferred from
astronomical
observations that this
is the case, because

FIG. 20 STREAMLINES FOR
DIFFUSING MEDIATORS

stars would appear very fuzzy, especially in the wake of
the Earth, if a turbulent boundary layer were to be
present. It follows that if space does behave as a
fluid, then it must be superfluid.

For ideal flow past a sphere the maximum streamline
flow velocity is 1.5 times the undisturbed velocity
measured far upstream. The corresponding velocity ratio
for a cylinder placed normal to the flow is 2. The
latter value is needed to explain stellar aberration, as
will be seen in due course. It is possible that space,
being composed of virtual particles, might behave
somewhat differently from normal fluids, so that the
two-dimensional pattern transformed to three, in the
known manner of ideal fluid flow, is appropriate. But
there is a second possibility.

With the bulk flow of gross space forming

streamlines about a near-spherical space bubble, hydrodynamic pressures will be imposed. There will be two stagnation points, one facing forward, the other rearward, and making the flow pattern symmetrical. Clearly, space bubbles will be flexible, being themselves fluid, and so they will be subject to distortion by hydrodynamic forces. They will be pinched across an axis joining the stagnation points and sucked out to greater radius in a direction perpendicular to this axis, - at the "rim" - owing to the need for the flow to accelerate as it passes around the bubble. For a sphere the velocity "v_r" at the rim is "1.5.u" , where "u" is the Earth orbital speed. The speed of the undisturbed fluid far upstream is equal to "-u".

As a result of these hydrodynamic forces the sphere will be distorted to the shape of an oblate spheroid and the local speed at the rim will be increased in consequence. A rough estimate using the ideal source and sink method has shown that a ratio of rim radius "R_r" to stagnation radius "R_o" of about 1.4 is needed to give a value of "v_r" at the rim of "2.u". This is the condition needed to fully explain stellar aberration, whilst allowing light to be propagated at a fixed speed relative to local space.

The flow pattern for gross space over such a bubble is illustrated in FIG.21 and contrasts sharply with the mediator diffusion pattern given in FIG.20.

2.7 AN EXPLANATION FOR STELLAR ABERRATION

Light from a star approaching from a direction perpendicular to the orbital motion of the Earth will pass through a region of highest stream velocity. A different frame of reference can be specified with Sun-space at rest and with the tangential direction parallel to that of Earth motion. Then at the bubble rim the fluid will experience a transient acceleration to a speed "v" equal and opposite the orbital speed "u" of the Earth and will cause light to be deflected backwards.

A model for the electromagnetic wave was introduced

in Chapter T.S.1 and
was illustrated in
FIG.9.* The way the
transverse vibrational
motion can be explained
was touched upon, but
for present purposes it
is not necessary to
take this into account.
Briefly the photon
jumps instantly from
one virtual electron of
space to another, being
repeatedly absorbed and
re-emitted. A sideways
movement of electric
charge occurs at each
absorption, so enabling
the neutral photon to
exhibit electromagnetic
properties. This
motion can be ignored
for present purposes
because it is
superimposed on the

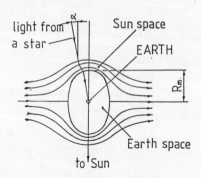

FIG.21 SPACE BUBBLE   Earth at rest

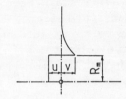

FIG.22   VELOCITY DISTRIBUTION
Sun frame at rest

effects of interest. Hence, for the present the photon
can be imagined as moving in a straight line as it jumps
between virtual electrons. Energy and momentum are
conserved by the photon at each step. A photon
travelling perpendicular to Earth orbital motion will be
carried backwards relative to Earth with speed "v" at
the point where the bubble boundary is reached, as shown
in both FIG.21 and FIG.22. Before crossing the sharp
boundary of the bubble the photon will carry a
tangential component of momentum "p_v", i.e:-

$$p_v \;=\; \frac{h}{\lambda} \cdot \left(-\frac{|v|}{c} \right)$$

This is due to the backward velocity "v" of the last
virtual electron, (carried along with the space-fluid)
from which it was emitted. But the next virtual

* See Page 269

electron which absorbs it will be inside the boundary, imparting momentum "p_u", i.e:-

$$p_u = \frac{h}{\lambda} \cdot \left(\frac{u}{c} \right)$$

So on the next re-emission the tangential momentum carried by the photon will be the sum of "p_v" and "p_u" and becomes:-

$$p = \frac{h}{\lambda} \cdot \left(\frac{u}{c} - \frac{|v|}{c} \right) \qquad\qquad [2.19]$$

This will have a zero value when "$|v| = u$". Hence the photon will again move in a perpendicular direction in the Sun-space frame. But now moving in Earth-space having velocity "u" such photons will appear to move, relative to Earth, with an angle of aberration "α" equal to "u/c". In this frame "$v_r = 2.u$" where "u" is the approach velocity far upstream. Taking Earth orbital speed "u" as 29.8 km/s, "α" becomes 10^{-4} radians or 20.3 arcseconds. This accords with observations made on the North star "Polaris" or any other star in a direction perpendicular to Earth orbital motion.

The path of a light ray is shown in FIG.21 and the transient velocity distribution through space at the bubble rim as the Earth moves past is illustrated in FIG.22.

Such a model then allows the electromagnetic wave to travel at a speed "c" relative to the space medium and yet satisfies both the Michelson-Morley experiment and stellar aberration. Space is at one and the same time dragged along with the Earth and yet the Earth still moves through space.

The model is also consistent with the observations made on eclipsing binary stars. This is so because light is propagating relative to local space and the rays from each member of any binary appear so close to one another that light rays from both pass through the same space for almost the entire distance. Hence the time taken to reach us is the same for both members to a high order of accuracy.

The solution depends, however on the ability of light to follow the velocity variations of space.

2.7.1   SHOWING THAT LIGHT WILL PROPAGATE
RELATIVE TO LOCAL SPACE

The argument assumes light to come into equilibrium with the motion of space both outside and inside the space bubble, so that it propagates relative to local space. It is readily argued that this will be the case. Between any pair of streamlines a band of fluid is to be assumed in which the flow velocity is constant. In regions close to the surface of a space bubble such bands of flow will differ in velocity from one to another. Velocity discontinuities need therefore to be imagined at the boundaries with adjacent bands. Photons jumping across these shear surfaces would be affected just as at the surface of the bubble, adding to themselves the average velocity differences between adjacent bands. They will not be in equilibrium with the band they enter. However, each band can be subdivided into a large number of narrower bands, all having the same velocity, and so, as the photon crosses from sub-band to sub-band, it comes closer to equilibrium with the band velocity. By the time it reaches the next shear boundary it will be in equilibrium. In the real case the shear surfaces will be so distributed that they will no longer really exist, but the argument shows that the photon will always propagate at a speed relative to local space, except when crossing a sharp boundary of real shear. This will only occur at the bubble surface.

In the full book(212) a further problem of importance is dealt with. This concerns the behaviour of superfluid compressible space consisting of balanced virtual particles of positive and negative energy. Such space can only have zero net inertia, yet it is shown that some separation of the two components is likely when space flows round bends. It is shown that an inertial amplification can arise to account for space bubble distortion to an oblate spheroid, as required to

provide the surface velocities needed for a full
explanation of stellar aberration. No further
discussion of this aspect will be made in this book.

2.8 COULD SPACE BE PARTLY PHOTON-CONTROLLED?

An essential part of the explanation given in this
chapter for stellar aberration was a model for the
electromagnetic wave as a structure, involving photons
coupling with the virtual electrons of space. The
immense energy density of the two halves of space
cancel, so that the net inertia of space, being
proportional to its net energy, is practically zero.
Hence any coupling by the most minute force should be
able to control or move local space.

The intensity of solar radiation "I_R" at the
distance of the Earth is 1,360 W/m^2. The time of travel
through a cubic metre of space is "1/c". Hence the
radiation energy density "η_R" will be equal to "I_R/c".
This works out to $4.5.10^{-6}$ J/m^3.

Space will also be controlled by the primary
mediators of gravitation, which have been termed
"exnegons". As the range of gravitation is assumed
changed from 10^8 to 10^9 light-years, the ratio of photon
to Earth surface exnegon energy densities "η_R/η_x" changes
from 6.10^7 to 6.10^9.

On the basis of relative energy density of photons
to exnegons it is the former which should have the
dominant effect as far as the induced motion of space is
concerned. Since photons are controlled by abstract
wave-functions they can jump from one virtual electron
to another. They could do this without affecting the
compression of space on which the force of gravitation
is dependent, because they would not diffuse through
space. The exnegons had to diffuse as described in
Chapter 5 and this caused the virtual particles of space
to be pulled toward one another. It is possible that
since photons must be coupled with space, momentum
carried by them could cause a bodily motion superimposed
on that which gravity creates.

On the other hand photons only interact with virtual

electrons of positive energy.   An exclusion principle
had to be postulated in Chapter 5 to prevent electrons
falling into negative energy states.   It could be that
the induced motion in the positive half is so small that
the negative half is not pulled along.   So in this case
the huge positive energy density of space is effective.
Then the fact of the net inertia being zero is
irrelevant.   This seems the most likely proposition.

The situation is by no means clear, however, so a
large question-mark has appeared.   An answer could only
be found by experiment but some speculation could help
in order to formulate ideas for such tests.

It is possible, for example, that photons caused by
heat radiation inside an aircraft cabin could cause the
space inside to be carried along, even though the
gravitational effects of the cabin would be too weak to
exert such influence.   This possible complication needs
to be borne in mind, though it will be ignored in the
next section.

2.9   SPACE BUBBLES AND MODIFIED RELATIVITY

A picture emerges in which the motion of objects
needs to be measured from a frame of reference fixed
with respect to local space.   The Earth will carry along
a space bubble which will move like a solid object
through fluid space of the Sun.   It will be the shape of
an oblate spheroid of mean diameter 260.10^6 metres, with
its rim stretching out to about 90% of the distance to
the moon.   The moon will be contained in its own bubble
24.10^6 metres in mean diameter.   In turn the bubbles
around Sun and stars will travel like solid objects
through the fluid space of the galaxy and so on.

Not all objects would be surrounded by space
bubbles, because a critical mass exists.   As the mass is
reduced and assuming for the sake of simplicity that the
shape remains spherical and density constant, the bubble
radius will shrink faster than the radius of the object.
It is readily shown from equation[2.18] that in the
field of the earth alone and at a distance R_E from its
centre an object of surface radius "r_s" and matter

density "ρ_b" will have a space bubble radius "R",
before hydrodynamic distortion, given by:-

$$\left(\frac{R}{r_s}\right)^2 = \frac{4}{3}.\pi.r_s.\rho_b.\frac{R_E^2}{m_E} \qquad [2.20]$$

Hence a critical radius will be reached when "R = r_s,
because at the surface the exnegon flux from the surface
of an object is a maximum. Therefore, on reaching this
condition the bubble will collapse altogether and the
object will not be in control of any local space, other
than that provided by any possible photon coupling.
Hence small objects will move with local space flowing
through them.

It is interesting to imagine a massive object like
the Earth, given a linear acceleration to a high speed
"u" from rest with respect to the average velocity of
remaining matter in the universe. The latter is to be
called the "zero frame" and would relate to a point at
the centre of the "Big Bang" of creation. Then a
kinetic energy "δE_u" will have been imparted to each
kilogramme of rest mass "m_o". If one kilogramme of rest
mass is then projected backwards at the same speed "u",
it will be at rest with respect to the zero frame. The
question is:-

"Will the energy now be back to the original value
of "$m_o.c^2$", i.e. "E_o" ?"

At rest with respect to the local space bubble, the
energy will be "E_1", equal to "$E_o + \delta E_u$". Then to shoot
the object backwards the energy will have to be
increased to "E_2" where:-

$$\frac{E_2}{E_1} = \frac{E_1}{E_o}: \quad So \quad \frac{E_2}{E_o} = \left(\frac{E_1}{E_o}\right)^2 \qquad [2.21]$$

There is an apparent contradiction, but this is
because the object is moving inside the Earth's space
bubble. As soon as it breaks out to enter the space
which is stationary with respect to the zero frame, an
energy interchange with space will create a rebalance

and the original energy "E_o" will be restored. The energy carried away inside the Earth bubble will then be the same as if this rejected mass had not been incorporated in the first place. The conservation laws are obeyed, as they need to be. The reduction in positive energy of the object will be exactly balanced by the destruction of some negative energy of space.

It follows that when calculating trajectories there will be a correct frame of reference for evaluating kinetic energy and this will assume local space as stationary. When a bubble boundary is breached to enter space moving with respect to the previous bubble, the correct frame of reference changes. Ultimately, by changing to larger and larger space bubbles a state could be reached in which velocities would be related to the zero frame. In this frame alone objects would be related to their true rest mass or rest energy. In "higher" frames local space would be moving relative to the zero frame at some speed "v". So the apparent rest energy on which all dynamic calculations need to be based would be "$[1 - (v/c)^2]^{-1/2}$" times greater than the true rest energy. It is necessary to consider the idea of "directed energy analysis" in order to develop a theory able to define the way of proceeding.

Energy cannot be regarded as a vector quantity, because it cannot be added in the manner defined for vectors. The components of energy "δE_x", "δE_y" and "δE_z" need to be added arithmetically to obtain the total kinetic energy "δE". However, energy can be directed in the sense that the kinetic energy of a bullet is directed energy. But kinetic energy can be split into components having different directions.

In the present case it is convenient to consider a small object within, say, the bubble of a large comet in highly elliptical orbit about the Sun. The object can have a velocity "q" relative to the comet. Then it will have an energy component "δE_p" parallel to comet orbital velocity "w" and one in the transverse direction "δE_t". These correspond with velocity components "q_p" parallel and "q_t" perpendicular to the relative wind of speed "w" of space past the comet, measured at a point far enough

upstream to eliminate the influence of streamlines of
space flowing around the bubble. The same criteria
would apply to flow of a fluid. The total enthalpy of
the fluid is constant at all points and is equal to the
value measured far upstream.

Then on breaking out of the bubble into Sun-space,
the coordinate system needs to be changed so that motion
is now measured relative to Sun-space. It follows that
a directed energy corresponding with "w", i.e."δE_w",
needs to be added to "δE_p" but nothing is added to "δE_t".
The addition rule can be defined by:-

$$\delta E_{ps} \;=\; \delta E_w \;+\; i.\delta E_p \qquad\qquad [2.22]$$

i = +1 if "q_p" is in the same direction as "w"
& i = -1 if "q_p" is opposite the direction of "w".
("w" is the relative wind)

The total kinetic energy of the particle in Sun-
space will then be obtained by the arithmetical addition
of "δE_t" to "δE_{ps}". The new direction of motion, relative
to the previous direction of "w" can be found from these
two energy components using the rule developed in FIG.11[*]
of Chapter T.S.1. The new speed relative to the Sun
reference frame will also be obtained during the
calculation. A corresponding reduction in apparent
rest-energy and mass will also occur during the
transition but this is not a true change. It is a
change in the kinetic energy originally added and
considered as effective rest-energy, owing to the
difference in speed of local space from that of the zero
frame.

The problem of dealing with the rest-energy is best
dealt with by first thinking of a reference value
corresponding with a position at the centre of the Sun.
A test object is first lifted on a cable at low speed
from this centre to the distance of the comet. There is
no change of rest-energy "E_o" during this process. Then
it is accelerated to comet speed "w" by adding energy
"δEw". The apparent rest energy of an object in the
comet bubble is now "$(E_o + \delta E_w)$". Just inside the

bubble, extra kinetic energy "$(\delta E_p + \delta E_t)$" is added to provide relative speed "q" and total energy inside the bubble rises to "$(E_o + \delta E_w) + (\delta E_p + \delta E_t)$" whatever the direction of motion of the test object relative to the comet.

When it breaks out and enters Sun-space the rest-energy falls back to "E_o" but a curious result emerges. When "i = +1" and the value of "δE_{ps}" from equation [2.22] is evaluated, the total energy is found to be exactly the same both inside and outside the bubble as is to be expected, the apparent extra rest-energy inside being transmuted to extra kinetic energy outside. But when "i = -1" the total energy outside is only "$E_o + (\delta E_w - \delta E_p + \delta E_t)$". Hence a drop, attributable to an energy interchange with space, of "$2.\delta E_p$" has occurred.

For low relativistic speeds these relations show that there will be negligible velocity change during transition, except for that involved in adding the velocity vector corresponding to change of reference frame. There will be a change in relative mass and energy corresponding with the change in frame of reference only for objects emerging from bubbles in a regressive direction. There will be a time dilation effect associated with this energy change and for small velocities can be found by applying a binomial expansion to equation [2.9]. The result is "$-(q_p/c)^2$". Hence if a space probe emerged with a regressive velocity at Earth's space bubble of, say, 10,000 m/s, then a sudden time dilation of amount equal to -1.1 nanoseconds per second should occur. This would be reflected as a small increase of transmission frequency in any radio signal from the probe equal to 1.1 parts per billion.

This would be easily measurable by modern atomic clocks and so a very useful check on the theory could be made by making use of Venus bound probes. It would be possible to measure the shape of the rearward face of ths Earth's space bubble by this means but outgoing probes could not sense the forward half. But regressive velocities can be experienced by incoming probes and

these could be used to locate the forward facing part.
In this case a frequency drop will occur as the probe
enters.

The previous analysis applies, of course, only to
particles having finite rest-energy in the zero frame.
For particles like photons having no rest-energy in any
frame, the effect of crossing a bubble surface will have
a different effect. Since light propagates relative to
local space its absolute speed relative to the zero
frame will change on entering a moving bubble. In
consequence a shift in relative energy and wave
frequency will arise for its matter-waves in accord with
equation[1.22]. There is no reason why this effect
should correspond with the frequency change for radio
transmission just descibed, since this is locked on to
the frequency of mechanically governed oscillation.

The change in relative mass or energy is caused by
interchange with the quantum vacuum. This change to a
new mass will be real, however, and will be the same
whatever the relative speed of the observer, a
conclusion at variance with that of special relativity.
Relative mass varies with speed of the observer in this
accepted theory and constitutes another anomaly. An
object can only have a single value of mass or energy;
it cannot in principle be a variable with respect to the
relative speed of the observer.

According to the modified view of special relativity
there will be an absolute frame of reference, after all,
at which objects at rest will have their true rest mass.
If the universe arose in the "Big Bang", then this would
correspond with its centre. There is little hope of ever
establishing this absolute frame because it is not
possible to view the entire universe. This does not
mean such a frame is non-existent, however.

The modified form of relativity described enables
light to be propagated relative to local space. It
differs in many respects from special relativity but
should not be dismissed, because it does not contain any
of the internal inconsistencies from which Einstein's
theory suffers. It is not enough that a theory is able
to give accurate predictions, confirmed by experiment;

it must also be free from internal contradiction.   The
new theory has this advantage but new experimental
checks need to be explored for purposes of verification.

2.10   NEW EXPERIMENTAL CHECKS (1 - 7)

1)   The cheapest check has just been suggested.   It
would measure the sudden time dilation predicted to
arise when an object moves out of one space bubble into
another in an orbital-regressive direction.

   If no sudden dilation arose, then Einstein's special
relativity would appear to be supported despite its
internal contradictions.   But because of these the
assumptions of special relativity cannot be regarded as
correct.   So if the modified proposals suggested here do
not satisfy the experimental checks, it would be
necessary to continue the search for a better concept.
However, the lack of internal contradiction in the new
theory suggests that a positive result is most likely to
be observed.

   This test could be carried out very cheaply by
making use of the next space missions which involve
visits to the inner planets, Venus or Mercury.   The only
extra equipment needed would be an Earthbound radio
receiver equipped with an oscillator tuned to an
existing transmitter aboard the spacecraft.   A beat
frequency should arise due to the expected sudden
frequency shift as the Earth space bubble boundary is
penetrated.   Indeed the minute effect may already have
been observed but dismissed as an electrical aberration.
Unless a researcher is unusually observant, unexplained
effects tend to go unnoticed and are not reported.   They
are observed when a new theory appears and they are
searched for as a deliberate act.

   It is worth noting at this point that although a
frequency jump should occur at the bubble boundary,
there will be no change in the speed of light relative
to space.   But space is moving at a different speed
outside the Earth space bubble, consequently the speed
of light relative to Earth will be altered.   This poses
a new question.   "Will the Shapiro time delay be

affected?" In general the answer has to be "Yes" for
elliptical orbits, but by an undetectably small amount.
Observed from the Sun space frame, light would travel at
the same speed as measured inside the Earth bubble at
equal gravitational potentials. The majority of the
distance is travelled in Sun-space, so corrections for
the speed change inside the bubbles would be very small.
For near circular orbits like those of Earth and Mars,
the velocity shift is in any case perpendicular to the
line of sight when Mars is close to superior
conjunction, in which case no discrepancy can arise.

2) From the arguments presented it is also clear that
efforts need to be made to check the prediction that
light is propagated relative to local space.
Measurements of the speed of light in different
directions are therefore required, using moving frames
of reference.

A Michelson-Morley interferometer could be used for
this purpose but fixed to a moving platform. A very
suitable mounting would be the cabin floor of a Jumbo
jet, because such aircraft fly very smoothly. The test
could be carried out quite cheaply during a normal
commercial flight. If a null result is returned, it
could be interpreted as being due to the photon coupling
of space to the cabin walls. Space inside the cabin
might be dragged along.

3) The third experiment would be made in an Earthbound
     laboratory and would repeat the Michelson-Morley
experiment *in vacuo* but in motion relative to the Earth.
This could be carried out, for example, with the
apparatus mounted on the end of a rotating arm.
Corrections would need to be made for distortions caused
by centrifugal forces. It would not be an easy
experiment to mount.

But a better way would use a hollow evacuated
revolving drum about 0.3 metres in diameter and some 1
centimetre long inside. It would be fitted with two
optically flat windows. These would both be arranged in
planes passing through the axis of rotation and fitting

into a notch-shaped cut-out at the drum periphery occupying a sector of about 30 degrees. The cylindrical internal surface of the drum would be made optically reflecting, so that light-beams entering one window would make almost a complete circuit to emerge through the other window. An interferometer could be arranged fixed to the laboratory floor and outside the drum. A light-beam split by a half-silvered mirror would then be arranged to pass through the drum in both directions simultaneously, one with the direction of rotation and the other in the counter direction. The beams would be recombined outside the drum to form an interference pattern. Care would need to be taken to ensure the intensity of the beam low in comparison with the black body energy density of radiation inside the drum, otherwise the test beams could control the space through which they passed and so cause a null result to be returned. It is not impossible that radiation from the molecules of any gas present could provide space coupling, so that space would follow movements of the gas. This is why a vacuum environment is necessary.

Properly executed, interference fringe movements ought to be observed if heat rays couple virtual electrons of the internal space to the walls of the rotating drum. If space is only controlled by the dominant gravitational field of the Earth, however, a null result would be returned. But in this case one would expect a positive result from the test made in the Jumbo jet, described as experiment No.2.

The same apparatus could be used filled with liquids and gases to determine the degree of coupling induced by matter over a wide range of densities down to pressures close to the high vacuum state. This would be a similar test to that described by Einstein(110) and carried out by Fizeau. He only used water flowing through a "U" bend, however. A rotating drum would be much more accurate owing to the elimination of errors caused by the non-uniform velocity profiles which are always present in flowing liquids.

Experiments 2) and 3) together would therefore determine whether or not photon coupling could cause

local space to be dragged along with an object. If both
flying and rotating drum tests returned null results,
then the present assumptions and the associated modified
special relativity would need revision.

4)   The time dilation experiments already carried out to
check the predictions of special relativity also support
the modified version described here. Consequently a new
version needs to be designed capable of discriminating
between them.   So far only moving clocks have been
observed from the ground and compared with ground-based
clocks. If some checks had been made from the flying
clocks, however, the modified theory suggests that
different results would have appeared.
     A relatively cheap experiment as compared with any
in orbit about the Earth would involve the installation
of two atomic clocks, one in each of a pair of Concordes
flying at twice the speed of sound. The clocks keep
perfect time with one another when the aircraft are
flying side by side in the same direction. But then
both aircraft are made to circle in opposite directions
away from one another but maintain the same height and
speed.
     According to Einstein, when the planes are flying
apart or approaching head-on, each clock will see the
other running slow, after correcting for Doppler shift.
This is a consequence of the assumption that any frame
of reference sees light-beams moving at a universal
fixed speed which is not related to any medium.
     In the modified theory time dilation is due to the
increased inertia of elementary particles resulting from
an increase in kinetic energy relative to local space.
The atomic clocks would measure no dilation whatever the
directions of relative motion, provided both planes
maintained equal airspeeds and flew at the same height.
Furthermore, if a ground-based clock were also included,
Einstein would predict the ground clock to be running
slow as seen from either airborne clock, whereas the
modified theory would predict it to run fast.   The
latter prediction is consistent with the commonsense
approach, since it is consistent with observations which

have already been made from the ground. It is assumed
that in addition to correcting for Doppler shift the
results are corrected for the extra time dilation due to
height, the gravitational red shift.   As shown in
Chapter T.S.1, this correction is the same for both
theories of gravitation.

In 1976 R. Vessot and his team launched a Scout
rocket on a two-hour sub-orbital flight as beautifully
described by Will(124).   It contained a hydrogen maser
clock operating at a frequency of 1,420 megahertz.   An
identical clock was earthbound. Very clever electronics
were incorporated which cancelled out the Doppler shift,
whose magnitude dwarfed that of the relativistic effects
which the experiment was designed to check.

The experiment verified the time dilation and
gravitational red shifts, with an accuracy of 70 parts
per million, which Einstein's theory predicted. Hence
this trial also supports the extended Newtonian physics
to exactly the same degree.

More important for the present proposal, is the
demonstration that apparatus already exists having
adequate resolution.   The methods Vessot used gave a
resolution of 1 part in 10^{12}, whilst Einstein's theory
for the Concordes flying apart predicts a dilation of 9
parts in 10^{12}. This resolution is not as good as would
be desired but it would be adequate for discriminating
between the two theories.

5)   Tests in orbit about the Earth would produce the
most convincing results, but because of the expense
involved they would only be mounted if a reasonable
chance of success could be expected. All possible
Earthbound tests therefore need to be explored first.

It would be well worthwhile to carry out a cheap
investigation by analysing existing astronomical data.

According to Møller(210) the theoretical equation
for stellar aberration based on special relativity is:-

$$\tan(\theta') = \frac{\sin(\theta)}{\cos(\theta) + \dfrac{u}{c}} \qquad [2.23]$$

Here "θ" is the angle between the direction of motion of the Earth of orbital velocity "u" and the actual direction of the star, whilst "θ'" is the corresponding apparent direction. He says this equation is in complete agreement with observations but does not state to what degree of accuracy. An oblate space bubble would give a very similar result but an exact theoretical treatment ought to show a very slight discrepancy of the order 1 arcsecond at a rough guess, for angles in the region of 45 degrees. When "θ" is close to 0 or $\pi/2$ discrepancies with respect to equation[2.18] ought to be close to zero, whether based on special relativity or the new approach.

In a private communication Professor Archie Roy of Glasgow University stated that he thought an accuracy of about 0.1 arcseconds ought to be possible, so any discrepancies could be used to justify commitment to an orbital test. If the discrepancy was found to be exactly zero, the result could be interpreted as showing space bubbles to be spherical, so that there would be no hydrodynamic amplification by space and the alternative explanation for "$v_r = 2.u$" might be applicable.

6)   Only an experimental check using a spacecraft could really settle the question. The above analysis, together with the much more convincing evidence which experiment No.1 could provide, would be sufficient to justify the expense of an orbital mission. A small telescope of no more than 50 mm aperture but having a very high magnification would need to be gyro-stabilised and projected into a very elongated elliptical orbit about the Earth. Its apogee would be about three times the distance of the moon. The orbit of the probe would remain fixed in space as the Earth travels in its own orbit, making possible the mapping of the shape of the space bubble of the Earth, including perturbations

caused by the moon. In this way the very existence of
the quantum vacuum of space might be confirmed. The
experiment is expected to show that space behaves as a
compressible superfluid.

The probe would preferably measure the aberration of
Polaris. The North star is chosen because the line of
sight will be least influenced by the effect of the
moon, surrounded by its own space bubble. This is of
such a size that it can just merge with that of the
Earth. The combined bubble would have a lump on one
side, making it asymmetric. For the same reason the
first probe would best penetrate the bubble surface near
to the North or South. A later probe could orbit in the
plane of the ecliptic in order to measure moon bubble
effects.

There should be no change in the observed aberration
until the edge of the Earth bubble is reached; then a
sudden extra deflection of 20.3 arcseconds should occur,
doubling the normal aberration reading. Gradually the
deflection should decay to normal as the region of
accelerated flow around the Earth's space bubble is
traversed.

Such a probe ought also to include an interferometer
on the Michelson-Morley pattern. It should demonstrate
a measured speed of light equal to "$(c - u)$" in the
forward direction and "$(c + u)$" measured backwards,
where "u" is now the speed of the probe relative to
Earth-space. An external mounting might avoid possible
interference by photon control of space.

If null results were consistently returned, despite
the use of photodetectors to permit weak beams, it would
have to be conceded that light did not propagate
relative to local space.

7) The seventh experimental check has already been
described fully in Chapter 11. It does not aim to test
the modification to special relativity. Instead this is
designed to discriminate between the extended Newtonian
theory of gravitation and general relativity. It is so
far the only new check on gravitational theory to spin-
off from the new concept. This is a laboratory

experiment aimed at showing the existence of negative
energy states and to show that gravitational potential
energy is only a pseudo-energy form. If negative energy
states exist, then it is feasible for positive and
negative states to arise or vanish together in equal and
opposite amounts.   Pure creation or destruction of
energy is permissible under these circumstances without
contravening the laws of conservation of either energy
or momentum. The experiment should demonstrate that it
is feasible in principle though on only a minute scale,
to both create and destroy energy. If a positive result
were returned, this check would support the extended
Newtonian physics by showing that gravitation involves
a real force. Because general relativity is based on
the concept that objects in free fall move along
geodesics in curved space-time without involving any
force, it follows that a positive result would
definitely show Einstein's concept to be false.

This would be the most exciting experiment of all.
It could open up a new avenue of research to discover a
new and totally non-polluting source of energy.

2.11   CONCLUSIONS TO CHAPTER T.S.2

It is concluded that special relativity can be
modified in order to eliminate its internal
contradictions. The three main equations which relate
relative mass to rest mass, energy to mass and predict
the relativistic time dilation can be derived by
alternative means which consider atoms being
accelerated. The way light propagates is irrelevant to
the derivation.   But it followed that no Lorentz
contraction was possible.   According to Einstein,
objects would distort on speeding up becoming shorter
but remaining unchanged in the transverse direction.
Objects were shown to remain of constant shape when
accelerated, even up to the speed of light. This could
only be interpreted to mean that Einstein's initial
assumption, stating that the speed of light was the same
whatever the frame of reference of the observer but

propagating independently of any medium, was unsupportable.

An alternative deduction that light propagated relative to local space led to predictions which were entirely consistent. A new explanation was then required, however, to show how the observation of stellar aberration could be reconciled with the Michelson-Morley null experiment and the observations of eclipsing binary stars.

The conceptual difficulties of the theory of special relativity were shown to derive from unacceptable logical inconsistencies.

Firstly, the contraction in the direction of motion is a direct prediction from Einstein's initial assumption. An observer moving with the object, however, sees no contraction in length. How can an object change in length in one frame of reference and not in another? The acceleration treatment showed such contraction to be impossible.

Secondly, Einstein's theory states that an observer moving with the object sees only the rest mass, whilst the relative mass, which is higher, is seen by the stationary observer. It is impossible for the true mass to vary with the velocity of the observer. If an object is accelerated, then energy and therefore mass is added and the increased mass will be seen by both moving and stationary observers. The slowing of time then becomes explicable because of increased inertia and a reduction in the abstract force which maintains the size of atoms constant. The moving observer will be similarly affected. Hence the moving observer would think the mass remained at its rest value if he used moving matter as a basis of comparison. But this is not the sense in which Einstein's theory says that the rest mass is observed.

The difference can best be explained by thinking of a spring balance. With mass measured by the force required to give a certain acceleration, recorded as a deflection of the spring balance, the moving observer will not be deluded. Then an observer on the moving object would measure only the rest mass according to

Einstein. But the higher relative mass, equal to that measured by the stationary observer, would also be recorded by the moving observer according to the new theory. The Einsteinian prediction contains an internal contradiction which is not present in the new theory.

Thirdly, this leads to another logical inconsistency in special relativity which does not arise in the modified version proposed here. Two objects "P" and "Q" having equal speeds can be imagined. If they move in the same direction, then they both see the other as having only the rest mass "m_o" despite the energy added by acceleration to speed "v", and both see time moving at the same rate, according to Einstein.

If they have oppositely-directed motions, each sees the other moving with a relative speed much higher than "v", which is perfectly reasonable. But "P" sees "Q" as having a mass "m" higher than itself and operating in dilated time (correction for Doppler effect being assumed). Hence it follows that "Q" must see "P" as of smaller mass and vibrating in quicker time. This is again an accepted inconsistency, because in special relativity, "Q" sees "P" exactly as "P" sees "Q", which conflicts with the previous deduction.

Fourthly comes the puzzle of non-simultaneity. Events occurring simultaneously at places equidistant from both moving and stationary observers must be observed as simultaneous by both and this was shown to be so in the new theory. Special relativity, however, predicts a contrary result.

It is these inconsistencies which make special relativity seem so difficult to comprehend. The student is expected to take them aboard as if they were logical deductions and this is impossible. They make the theory impossible to accept.

Finally we come to difficulties with the propagation of light. In the theory of quantum gravitation developed here light had to propagate relative to local space. Only when this was accepted could a plausible structure be deduced, able to account for electromagnetic properties. [Dealt with more fully in Reference (212)]. This structure, combined with space treated as a superfluid, enabled an alternative explanation for stellar aberration to be advanced which was also consistent with the Michelson-Morley experiment and observations on eclipsing binary stars. The Earth was shown to drag along with itself a space bubble which would tend to be of spherical shape. Sun-space would flow over such a bubble with a maximum stream velocity of 1.5 times the approach velocity, whereas a factor of 2 was needed to fully account for stellar aberration. It was shown, however that hydrodynamic pressures would tend to distort the shape from a sphere to that of an oblate spheroid. This could provide the correct doubling of aberration as required. It is true that the explanation depends on an amplification factor for the hydrodynamic pressures of space being a certain value of about 10^7, but in (212) it is shown that space could produce this amplification.

But at least this chapter shows that explanations can be found for stellar aberration which do not depend on the initial assumption of special relativity and all the inconsistencies which then arise. The dependence on a coincidence, an amplification factor which just increases the maximum relative fluid space velocity from 1.5 to 2.0, is not unique to quantum theory. Indeed it abounds with strange coincidences. The working of the entire universe seems to depend on a match, to unbelievable exactness, for a surprising number of them.

However, such a condition is consistent with the theory advanced in the main text to explain wave-particle duality. An underlying unity has to exist in the universe which couples everything to everything else. The universe had to be created as a deliberate act by an all-pervading Grid possessing computer-like properties. The apparent strange coincidences then have another interpretation. They were deliberately contrived so that a system of interpenetrating universes could exist, capable of supporting intelligent forms of life! These were required to coexist as a number of interpenetrating systems of matter organised upon the common Grid.

The new approach provides a modified theory of

relativity which depends on space being compressed by
mediators emanating from ponderous masses. Local space
is locked to these and encloses them in a bubble over
which surrounding space is caused to flow. Motion needs
then to be related to local space which is regarded as
the "rest frame".

Seven new experimental checks were proposed. Some
were designed to find out how light propagated with
respect to space. Others were designed to discover
coupling effects of space to matter, to see whether
photons as well as the mediators of gravitation could
produce coupling effects. One of them aimed to test a
new prediction: that a time dilation would arise as the
boundary of a space bubble is crossed. Another would
aim to discriminate between Einstein's theory and the
extended Newtonian physics by testing differences of
predictions in regard to time dilation.

Difficulties would arise if such experiments all
returned null results. It would then be necessary to
find another explanation for stellar aberration and this
would be very difficult. Confirmation of the existence
of space bubbles by experiment is the next step needed
before much further progress can be justified. However,
the main theory of quantum gravitation would not be
discredited whatever the outcome of such experiments,
because it is consistent with equations[2.4] and [2.5]
usually attributed to special relativity. These are now
derived regardless of the properties of light by
considering acceleration alone.

Only one of the checks related directly to quantum
gravitation. This aimed to prove that a real force of
gravity existed. A positive result would establish the
new theory and at the same time invalidate any based
upon the concept of space-time curvature, because in
this class of theory no accelerating force is involved.

CHAPTER T.S.3

THE ELECTROMAGNETIC FORCE
AND THE ENERGY DENSITY OF SPACE

The force of magnetism has been related by physicists to
the electric force using abstract mathematical
reasoning. Only the electric force will be considered
here. A hypothesis has been worked out which can relate
these forces by a mechanical model and will be included
in the full technical book(212). It will not be
included in the following derivation. Here we are
primarily concerned with obtaining a value for the
energy density of space "ε", in order to show the degree
to which the present theory is able to relate the force
of gravity to the electric force. As argued in
Chapter 5, a buoyancy model for the electric force is
needed, since the alternative model based on mediator
absorption predicts an orbiting electron to continually
gain angular momentum.

The method to be used will derive an expression for
the force of repulsion between two electrons. The
measured force will then be used to determine the
required energy density of space "ε_s" as determined at
the mediator-emitting surfaces of the electron. This
will give the minimum value of "ε" for space as a whole,
because other virtual particles of space will add their
supplement. Also, the total energy density of space
must be uniform, like a gas, except for the compression
by gravity. Even at the surface of a neutron star,
however, the energy density of space can only be about
50% greater than at infinity. Hence for most
calculations "ε" can be considered uniform.

A value for "$\varepsilon_1/\varepsilon$" can be readily calculated using
equations [1.11], [1.12] and [1.35]. For example a

neutron star might be imagined to have the mass of the Sun, $1.99.10^{30}$kg, at the energy density of the neutron, $2.7.10^{34}$ J/m³, so that its surface radius works out at 11.65 km. Then "$\varepsilon_1/\varepsilon$" becomes 1.46.

The total value of "ε", considered as spatially uniform, will be the sum of partial values analogous to Dalton's law of partial pressures for a perfect gas. Indeed a direct relationship between energy density and pressure was shown to exist in Chapter T.S.1. The specialised mediators of the electric force will derive from unspecialised virtual particles bombarding their surfaces, changed by a catalytic effect and thrown off again. As they spread out from their source their partial pressure will diminish, so producing a partial pressure gradient which is extremely high as compared with the total pressure gradient induced by gravitation.

Indeed, for mediators moving at the speed of light it was shown in Chapter T.S.1 that the pressure "P" which they generate is equal to "$\varepsilon/6$". For massive particles the "P" is smaller but still proportional to "ε". Hence partial pressures and partial energy densities will vary in step with each other. Then as the partial pressure of mediators reduces with distance from an emitting particle, other virtual particles of space will fill in to make up the total pressure to a constant value. This model of space is illustrated later in FIG.24.* It is further to be assumed that the total energy and pressure impinging on the surface of any particle will equal reflected values. This represents the necessary equilibrium condition.

Hence the first requirement is to derive a plausible model for the electron. This will now be attempted.

3.1 MODELLING THE ELECTRON

Five properties of the electron are known and will be used to deduce a plausible model. These, taken from Blanchard(202) are:-
1) Electron net positive mass, "m_e" = $9.1091.10^{-31}$ kg
2) Electronic charge   "e" = $1.6021.10^{-19}$ coulomb

* See Page 356

3)   Apparent radius   "r" = 10^{-18} m
     (from collision interactions)
4)   The angular momentum of the electron, known as
its "intrinsic spin", "S", was deduced by Paul
Dirac and is:-
     S = $9.1322.10^{-35}$ N.m.s (to be termed "spin")
5)   "Mspin", the magnetic moment associated with
spin and given to sufficient accuracy by:-

$$M_{spin} = S.\frac{e}{m_e} \qquad [3.1]$$

As well as matching these properties a mechanism needs
to be provided for creating the spin and magnetic
moment. What maintains this spin?
     To begin the synthesis of a plausible model, the
electron will be considered as an elementary particle
arranged in the simplest possible way for mathematical
treatment. This is a circular hoop of radius "R" able
to spin about its polar axis.
Then if the rim speed is "v":-

$$S = m_e.v.R \qquad [3.2]$$

Only a minimum value for "R" will be calculated.
Clearly "v" will then equal "c". Inserting values this
yields:-
          "R" = $3.344.10^{-13}$ m

     Another value can be found from "Mspin" because
the current "i" in a circular orbit is equal to charge
"e" times orbital frequency and the magnetic moment of
a current loop is equal to the current times its
enclosed area. This yields:-

$$M_{spin} = 0.5.e.v.R = -S.\frac{e}{m_e} \qquad [3.3]$$

and (noting the -ve sign is only a direction
convention):-
     $-S.e/m_e$ = $-1.60617.10^{-23}$ coulomb m^2/s

If "v = c", then this gives a value of "R" equal to:-
    "R" = $6.6882*10^{-13}$ m

Immediately two problems arise:-
Firstly, to satisfy the magnetic moment the radius of
the hoop needs to be twice the radius needed to give the
correct value of spin. The conclusion must be that the
initial assumption made, that the electron is an
elementary particle, is false. To put matters right,
the mass of the hoop needs to be halved with the other
half of the mass placed at the axis of the hoop. But
now the electron is no longer elementary.

The second problem supports this conclusion. The
value of "R" must be the higher of the two, but this is
670,000 times greater than the apparent size from
collision interactions! It is also 610 times the size
of the nucleons with which collisions are made. It is
1.264% of the Bohr radius for the "H" atom.

It would be possible to reduce the hoop radius by
having two hoops in contra-rotation and with one
carrying positive charge whilst the other carried
negative charge. The negative charge would have to be
slightly in excess of the positive so that the
difference would represent "e". Also the counter-
rotating positive charge would need attaching to
negative mass so that the angular momenta of the two are
additive. The core particle can now be dispensed with
because a solution can be found without it. Indeed, if
"q_-" represents the charge on the hoop of positive mass
and "R" for each coaxial hoop is taken as the desired
10^{-18} m, the solution is readily shown to be:-

A ratio "q_-/e" = 334,411 with "q_+/e" one unit less:
and   "m_+/m_{e}" = 167,205 with "m_-/m_e" one unit less.

The solution still represents a compound particle,
however, consisting of both positive and negative mass.
The electrostatic force would pull the hoops strongly
together and would be reinforced by a magnetic

attraction.   Consequently, to maintain an axial
separation each hoop needs to produce a strong repulsive
force of very short range.   This can be achieved only by
the emission of heavy mediators from each to push
against the other.   These strong force mediators need to
have a mass sign opposite that of the emitting objects,
so that impinging mediators are of the same sign as the
masses pushed.   If they were absorbed, the hoops would
be slowed by the interchange of angular momentum.   It
follows that buoyancy-type repulsive forces need to
apply.   Conversely, a relatively small number of
mediators of opposite mass sign need to be absorbed in
order to create and maintain spin.   This mechanism will
be described, however, in more detail in the next model
to be considered.   When worked out in detail, the model
yields a universal energy density for elementary
particles "ε_p" which is about ten times the average for
space.

   This proposal will not be considered further because
the huge amounts of charge necessary seem impractically
large and the component masses seem unacceptably huge.
Only a subjective rejection is involved here because the
model fits the facts without any internal conflict, so
it could be correct.

   However, an alternative solution to be considered
next may seem more acceptable.

3.2 A PLANETARY ELECTRON MODEL

From the quark theory of nucleons it appears that unit
charge is "e/3", not "e" as at first assumed.   Symmetry
then suggests the electron to have its charge arranged
as three units.   Hence the hoop can be collapsed into
three planets, each of mass "$m_e/6$" and all orbiting at
radius "R" = $6.6882.10^{-13}$ m around a single core particle
of mass "$m_e/2$".

   The electron only has to *behave* as if it were a
particle of radius 10^{-18} m.   If each of its four
components is about this size, then such sub-units would
behave in a manner supporting observation during

collisions. In this way the large diameter of the orbit becomes acceptable and it is no longer necessary to involve huge and almost cancelling charge and mass.

The model is illustrated in FIG.23.

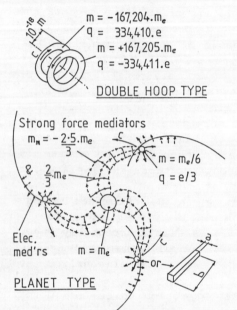

$m = -167,204.m_e$
$q = 334,410.e$
$m = +167,205.m_e$
$q = -334,411.e$

DOUBLE HOOP TYPE

Strong force mediators
$m_m = -\dfrac{2 \cdot 5}{3}.m_e$

$\dfrac{2}{3}.m_e$

$m = m_e/6$

$q = e/3$

Elec. med'rs $m = m_e$

PLANET TYPE

FIG.23 ELECTRON MODELS

The core particle emits mediators of negative mass radially outward to be absorbed by the planets in order to provide the necessary centripetal force for keeping them in orbit. All the charge is carried by the planets and is produced by the emission of virtual photons equally in all directions. These are of exactly balanced positive and negative mass so that no angular or linear net momentum is imparted by their generation.

Virtual photons of positive mass are generated with negative spin and virtual photons of negative mass have positive spin. This arrangement is consistent with the spin-coupling scheme given in FIG.6 of the non-technical section. It provides a means for like charges repelling whilst unlike attract. In this model a particle has an apparent volume with respect to recognised spin which is greater in ratio "f_v" to the volume when unrecognised. Consequently the values of "ε" to be deduced really need dividing by "$(1 - f_v)$", but this factor is unknown. For the specialised mediators of the electric force it can be expected to be much higher than for the gravitational

case.   Therefore it could be as high as say 0.5 or more
but will be assumed unity.

It is now necessary to explain how the electronic
spin can be produced. Let us suppose that the planets
are initially made as massive objects having low orbital
speed "v".   Mediators from the core particle, as
required for producing an attractive force on the
planets, when moving radially outward will appear from
the planets to arrive with an additional tangential
velocity component "-v" and so being of negative mass
"$-m_m$" will each add an angular momentum of
"$(-m_m).(-v).R$" which is positive. Hence the planets
will accelerate indefinitely. However the mediators are
virtual and so after absorption and decay no energy gain
is permanently imparted. At first the mediator will be
absorbed but then its energy is subtracted as its
borrowed energy is withdrawn.   In this manner the
increase in kinetic energy has to derive from rest
energy of the accelerated particle whose total energy
then remains constant. The limiting condition is the
speed of light, when all rest energy has been removed.
In this way the initial assumption that the planets move
at the speed of light is supportable by deduction. A
mechanism has been provided for generating and
maintaining a spin having a fixed quantum value.

3.3   CENTRIPETAL FORCE FROM MEDIATORS

The energy components deduced with the help of
FIG.11 and given in equation [1.22] can be used to find
the centripetal force needed to maintain an object
having total energy "E" in a circular orbit of radius
"R". Alternatively Newton's law that force is equal to
the rate of change of momentum can be applied.

In the latter case, which is the simplest to adopt,
the momentum is based on the so-called "relativistic
mass" i.e "m".   Then it is the rate of change of
direction of the momentum vector, "p" which governs the
centripetal force for circular orbit. This is "$p.d\theta/dt$"
and "$d\theta/dt = v/r$".   It follows that the classical

equation for centripetal force results in which the rest
mass is replaced by relative mass and can be written:-

$$F = \frac{E}{R} \cdot \left(\frac{v}{c}\right)^2 = \frac{E_o}{R} \cdot \left(\frac{v}{c}\right)^2 \cdot \left[1 - \left(\frac{v}{c}\right)^2\right]^{-1/2}$$

................[3.4]

It needs to be remembered that "$E = E_o + \delta E$", being the
total energy of the rotating object. Also "$E_o = 0$" here
since $v/c = 1$ and $E = m_e.c^2/6$ per planet.
There will be an additional electrostatic force but this
is readily shown to be only 0.000708 of the force
needed to counter centripetal acceleration and can be
included as a constant "k_e" i.e.:-

$$k_e = 1.000708$$

Also another constant can be defined, assuming
mediators and sub-atomic particles to have similar
shapes represented by their average radii. This allows
for the increase in capture cross section due to their
combined radii as compared with the radius of the sub-
atomic particle on its own. This radius ratio is;-

$$k_r = 1 \text{ approximately.}$$

"F" is to be balanced by mediators each of mass "m_m"
emitted from the core at speed "c" because this gives
the   maximum   momentum.   The   flux   "j"   of
mediators/(second.metre^2) was shown from equation[1.25]
equal to "n.c/4", where "n" is the number density of
mediators per unit volume. It is easiest to think of
flat surfaces with incident and emitted "j" values in
balance. Then since the shape of the core particle is
not known, any reasonable shape can be assumed. It
simply has to be accepted that some uncertainty cannot
be avoided. A cube of side "a" is mathematically the
simplest to choose since it has flat surfaces.

The total rate of mediator emission from the core
then becomes:-

$$J_{tc} = 6.a^2.j = 1.5.a^2.n.c \quad ....[3.5]$$

And with a mass "$m_e/2$" and the energy density of all elementary particles already shown in Chapter T.S.1 to be a universal constant taken as "ε_p", this yields:-

$$J_{tc} = 1{\cdot}5 . n . c . \left[\frac{m_e . c^2}{2 . \varepsilon_p}\right]^{2/3} \qquad [3.6]$$

The first case to be considered will assume the core mediators thrown out uniformly in all directions so that only a minute fraction are intercepted by the mediators orbiting at radius "R". If they are spheres of radius "r_p", then the capture area can be taken as "$\pi . (r_p . k_r)^2$", where "k_r" is a ratio to allow for mediator size. "r_p" can be found from the mass "$m_e/6$" and "ε_p".

The radial attractive force produced by mediator absorption by a planet will be equal to its rate of absorption of momentum. And mediators will be assumed to move at speed "c". This can be equated to the inertia force from equation [3.4] i.e.:-

Absorption   Inertia

$$F = \frac{(m_m . c . J_{tc})}{4} . \left(\frac{r_p . k_r}{R}\right)^2 = \frac{k_e . 2 . m_e . c^2}{6 . R}$$

$$\dots\dots [3.7]$$

Then with "r_p" written in terms of "ε_p" and noting that:-

"$\varepsilon_s = n . m_m . c^2$", simple algebra finally yields:-

$$\varepsilon_s = 12{\cdot}19 . R . \frac{e^{4/3}}{k_r^2 . (m_e . c^2)^{1/3}} \qquad [3.8]$$

The equivalent spherical core particle would have the same surface area as the cube of side "a" initially assumed in order to produce the same total mediator emission. Then "$r_c = 0.691.a$". If now "r_c" is made 10^{-18}m to match collision data, then "ε_p" works out to $9.9.10^{39}$J/m^3 and "$r_p = 0.6934.r_c$". Then if "k_r" is unity, equivalent to ignoring the size of mediators, the energy density of space "ε_s" from equation [3.8] becomes $4.04 \cdot 10^{46}$ J/m³.

This cannot be regarded as satisfactory because even if mediators expire after a range equal to "R" with none to spare, their resulting mass is $-1.03 \cdot 10^{12}$ times the mass of the electron! The reason is the very poor efficiency of utilisation of mediators. The fraction utilised is only $0.75 \cdot 10^{-12}$ of the number generated at the core. The model cannot be regarded as impossible because if an exactly equal number of dummy mediators of positive mass are simultaneously emitted, then both their mutual inertia and gravitational effects will be zero.

The Designer of the universe, however, might not have been satisfied with this result and so could have sought ways of improving the utilisation. Now a parallel situation is found in the binding of quarks into nucleons. According to Paul Davies(106) there are primary gluons which bind the quarks and secondary gluons to bind the primaries. Symmetry seems to demand this structure for the electron. Hence the core particle now throws out primary gluons of negative mass and these also act as flying particle factories to throw out secondary gluons of positive mass sideways to cause the primaries to bunch together. The planets do the same, so that strings are effectively formed, which connect planets to the core. Now all primary gluons can be utilised.

The resulting expression for the energy density of space as measured at the surface of the core particle becomes:-

$$\varepsilon_s = 2 \cdot k_e \cdot 2^{2/3} \cdot (m_e \cdot c^2) \cdot \frac{e_p^{2/3}}{R} \qquad [3.9]$$

Now the same data yields "ε_s" = $3.1 \cdot 10^{34}$ J/m³ (negligible as compared with a value obtained later for charge).

It is now possible to work out the total amount of virtual negative mass associated with primary mediators. It is the product of number flux "j_s or $n_s \cdot c/4$", mediator mass "m_m" and time of flight "R/c". The result is a numerical value 2.538 times the electronic mass for any value of "ε_p" initially chosen. There would need to be a similar amount of positive mass tied up in secondary gluons.

We now have a reasonably satisfactory model but this value of "ε_s" takes no account of the mediators generated from the planets for creating the effect of electric charge.

3.4 THE ELECTRIC FORCE

An equation will first be developed to give the force produced between two well-separated virtual photon emitters. For reasons previously discussed these emitters will be considered to be made up with flat sides perpendicular to one another. So that the effect of their shape can be studied, they will be assumed to be prisms of square section of side "a" and length "b".

3.4.1 MEDIATOR PRODUCTION

With a number density "n_s" of virtual photons (VP's) close to the surfaces, a VP flux "j_{ps}" will arise so that the total number emitted "J_{TP}" can be written:-

$$J_{TP} = (4.a.b + 2.a^2).j_{ps}$$

and since:- $j_{ps} = n_s.c/4$:-

$$J_{TP} = a^2.\left(\frac{b}{a} + \frac{1}{2}\right).n_s.c \qquad [3.10]$$

Now at a distance "r" these VP's are spread out over an area equal to "$4.\pi.r^2$". Hence the flux "j" at any radius "r" will be:-

$$j = \frac{\left(\frac{b}{a} + \frac{1}{2}\right).a^2.n_s.c}{4.\pi.r^2} \qquad [3.11]$$

3.4.2 MEDIATORS DIFFUSING THROUGH SPACE

Buoyancy-type forces demand the diffusing motion of mediators through space. The rate of progression is

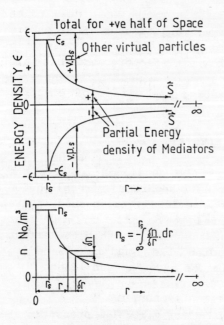

FIG. 24 PARTIAL MEDIATOR
DENSITIES OF SPACE

then proportional to their concentration gradient "dn/dr". In this respect they will differ from the motion of the virtual photons which established physics has assumed. This difference is justified, however, because the virtual photon absorption model had to be abandoned in order to match observation. Absorbers would gain angular momentum from attractive forces, as has already been shown, and this does not happen.

But at any radial distance "r" the number flux "j" has already been fixed by equation[3.11] and so this must be related with "dn/dr" by a "DIFFUSION COEFFICIENT", i.e. "K_d". Hence:-

$$\frac{\delta n}{\delta r} = -K_d \cdot j = -\frac{\left(\frac{b}{a} + \frac{1}{2}\right).a^2.n_s.c}{4.\pi.r^2}$$

. . . . [3.12]

Now the number density "n" must be zero at infinity but has the value "n_s" at the surface of some equivalent radius "r_s".

Another expression can be derived for the value of "n_s" by integrating "$\delta n/\delta r$" from infinity to "r_s". The relevant model of space is shown in FIG.24. Hopefully this will enable the unknown diffusion coefficient to be cancelled out.

The introduction of an equivalent radius is necessary so that this step can be completed. It does not matter if the equivalent sphere has a different particle density from the prism-shaped object assumed; all that is necessary is that it should have the same value of J_{TP} for an unchanged "j_s". Hence it must have the same surface area. Hence "r_s" is given by:-

$$r_s{}^2 = \frac{a^2}{\pi} \cdot \left(\frac{b}{a} + \frac{1}{2} \right) \qquad [3.13]$$

To proceed with the integration:-

$$n_s = -\int_\infty^{r_s} \frac{\partial n}{\partial r} \cdot dr$$

And "$\delta n/\delta r$ is given by equation[3.12]. Hence:-

$$n_s = \frac{K_d}{4 \cdot \pi} \cdot \left(\frac{b}{a} + \frac{1}{2} \right) \cdot a^2 \cdot n_s \cdot c \cdot \int_\infty^{r_s} \frac{dr}{r^2}$$

Which integrates to yield:-

$$n_s = \frac{K_d}{4 \cdot \pi \cdot r_s} \cdot \left(\frac{b}{a} + \frac{1}{2} \right) \cdot a^2 \cdot n_s \cdot c$$

Now "n_s" cancels because it occurs on both sides and "r_s" is given by equation[3.13]. Hence by substitution the required value of "K_d" becomes:-

$$K_d = \frac{4\sqrt{\pi}}{\sqrt{\left(\dfrac{b}{a} + \dfrac{1}{2} \right) \cdot a \cdot c}} \qquad [3.14]$$

This value for "K_d" can be substituted in equation[3.12] to provide an expression giving "$\delta n/\delta r$" which in the limit gives "dn/dr".

TECHNICAL SUPPLEMENT
3.4.3   THE BUOYANCY FORCE "F"

The buoyancy force is given by:-

$$F = V.\frac{dp}{dr} = \frac{V}{6}.\frac{d\epsilon}{dr} \quad \& \quad V = a^3.\frac{b}{a}$$

In addition for mediators of mass "m_m, since:-

$$\epsilon = m_m.c^2.n: \quad \frac{d\epsilon}{dr} = m_m.c^2.\frac{dn}{dr}$$

Hence:-

$$F = \frac{m_m.c^2.a^3}{6}\left(\frac{b}{a}\right).\frac{dn}{dr}$$

Combined with equations[3.12] and [3.14] an expression results giving the force between two prisms. But there are three per electron and so this needs to be squared i.e. multiplied by 9 to give the force of repulsion "F_e" between two electrons separated by a distance "D":-

$$F_e = \frac{3}{2.\sqrt{\pi}} \cdot \frac{b}{a} \cdot \sqrt{\frac{b}{a} + \frac{1}{2}} \cdot \frac{a^4.\epsilon_s}{D^2}$$

$$\dotso [3.15]$$

This can now be compared with the measured force given by:-

$$F_e = k.\left(\frac{e}{D}\right)^2$$

$$where \quad k = 8 \cdot 987.10^7 \ N.\left(\frac{m}{coulomb}\right)^2$$

Equated with equation[3.15] the result becomes:-

$$\epsilon_s = \frac{2.\sqrt{\pi}}{3} \cdot \frac{k.e^2}{\frac{b}{a}\cdot\sqrt{\left(\frac{b}{a} + \frac{1}{2}\right)}.a^4} \quad [3.16]$$

It is meaningful to rewrite this expression to eliminate

"a" by relating to the particle energy density "ϵ_p" instead.   Now:-

$$\frac{m_e \cdot c^2}{6} = \epsilon_p \cdot a^3 \cdot \frac{b}{a}$$

Hence, substituting in the above and substituting known data:-

$$\epsilon_s = \frac{8 \cdot 36 . 10^{-10} \cdot \epsilon_p{}^{4/3}}{\sqrt{\dfrac{b}{a} + \dfrac{1}{2}}} \cdot \left(\frac{b}{a}\right)^{1/3} \qquad [3.17]$$

In Chapter T.S.1 "ϵ_p" was taken as 2.10^{40} J/m^3. With the new refined model this should be halved, but the above value will be retained so that the following will be consistent with the gravitational analysis.   It only means that the core particle has been assumed 0.79 of the (very approximate) value given by Blanchard. Using this value the result is:-

b/a	a metres	b metres	ϵ_s J/m^3
1	$8.8.10^{-19}$	$8.8.10^{-19}$	$3.7.10^{44}$
100	$1.9.10^{-19}$	$1.9.10^{-17}$	$2.1.10^{44}$
10^3	$8.8.10^{-20}$	$8.8.10^{-17}$	$1.44.10^{44}$
10^4	$4.1.10^{-20}$	$4.1.10^{-16}$	$9.8.10^{43}$
10^6	$8.8.10^{-21}$	$8.8.10^{-15}$	$4.54.10^{43}$

From the above table it is clear that the effect of planet shape has only a small effect on the energy density requirement of space. It is, however, strongly dependent on the assumed average effective size of such particles. This can be judged by putting "b/a" = 1 in equation [3.16], since the effect of this ratio has been shown to be small. Then the energy density varies inversely as the fourth power of particle size. Unfortunately this size is only roughly known, so the above values could all easily be ten times too high.

3.4.4   A COMPACT HYBRID MODEL
Other models have been analysed. The most promising

of these was a hybrid arrangement in which the hoops of the first model are each replaced by three balls orbiting at the same radius as the hoop they replaced. Each had to throw off mediators of positive and negative energy as before in order to provide electric charge but, in addition, they had to throw off a type of strong force mediator to bind the balls together. These would not be able to couple with nuclear particles of course. It then turned out that a sufficent proportion of such "gluons" were intercepted so that these did not add significantly to "ε_s" because the balls had radii some 20% of the orbital value. A stable arrangement resulted when each ball threw off gluons of both positive and negative energy in equal amounts.

The average range of the gluons of opposite energy sign to the emitting particle needed to be very much shorter than the range of those emitted of similar energy sign. The chances of a particle decaying in a given time element were assumed equal for all gluons of a given type. An exponential rate of decay was then predicted. Forces had to be of the buoyancy type. Then energy "wells" arose which bound the assembly in a stable fashion.

At short range the high rate of gluon decay of opposite energy type created an extra component of partial "dp/dr" so providing stronger attraction than repulsion to balls of the same energy sign at very close range. This mutual attraction between all three balls of either ring had to be balanced by centrifugal force. The effect on the opposite ring of balls, made from the opposite type of energy, was to provide a long range attraction with short range repulsion, so maintaining the two rings at a safe distance so that their balls could not collide.

A small proportion of the gluons of opposite sign from the emitting particle had to be absorbed by the objects of opposite energy in order to generate and maintain opposed rotation of the two sets of balls. They automatically arranged themselves on a common axis but in two separated planes, so preventing collision.

Their spin angular momenta and magnetic moments then became additive whilst their mass and charge tended to cancel, as in the double hoop model, to provide the observed properties.

This model is mentioned because established physics suggests that the quarks making up nucleons have spins like that of the electron. If so they would also need a construction either of this or the double hoop kind first described in order to be small enough to fit inside the nucleus. They would have to be constructed of massive and almost balanced positive and negative energies like the hoop model. In this case the hybrid type would need to apply equally to the electron to provide a symmetry of construction for the atom as a whole.

Then our matter needs to be seen in a new light as an almost balanced system of positive and negative energy states. The residual positive energy of nucleons and electrons appears as only a minute fraction of the component energies.

Either type of compact model has to be associated with a higher energy density of space than the kind described with respect to FIG.23. This is because the planets, though of the same order of size in the region of 10^{-19}m, have to generate several hundred thousand times as much electrical charge of both kinds which then almost cancel. This increases "ε_s" to about 10^{50}J/m^3.

3.5   CONCLUSIONS TO CHAPTER T.S.3

The electron cannot be an elementary particle because its properties demand at least two components. A model which could account for its spin and magnetic moment and yet have an effective radius of only 10^{-18} m was described. It was a composite of two coaxial contra-rotating hoops. Unfortunately the composite needed huge and almost cancelling positive and negative electric charges equal to 334,000 times the electronic charge. This was at first judged to be a very unlikely option.

A probable representation was considered to be a

model having a central core particle accounting for half the electronic mass and having the remainder arranged symmetrically as three planets carrying all the charge and orbiting at a radius of 1.3% of the Bohr atomic radius. They had to orbit at the speed of light since the retaining mediators emitted from the core also maintained them at this limiting condition. The mediators had to be of two kinds. The primary kind needed to have a negative mass numerically equal to 2.54 of the total electronic mass and could be balanced by an equal amount of positive mediator mass. The latter had to act as binders for the primary mediators. Their contribution to the energy density of space was negligible.

A possible hybrid model consisting of two rings each of three balls, each ring of opposite mass, electric charge and sense of rotation was also considered. This had the same objection as the hoop model of requiring exraordinarily large component masses and charge but enabled quarks to be similarly modelled. It may be that these huge mass and charge values need to be accepted.

The mediators of the electric force required the energy density of space to be at least 10^{44} J/m^3. This represents each half of space as required to explain the electronic charge. But both halves would need increasing in the ratio 4/3 since the quarks of the proton are a mixture of 4/3 positive and 1/3 negative charge. Hence the gross density of space will be increased in the ratio 8/3, the halves cancelling their energies to leave a net energy of zero. The gross value is little different from that given by Starobinskii and Zel'dovich(215) of 10^{45} J/m^3, which they obtained by an unstated method. They regard this as all positive energy, however, and draw attention to the fact that some other factor must be present to make the net value zero.

Any figure between 10^{41} and 10^{45}, or even much higher, was shown in Chapter T.S.1 to be consistent with that required to account for the force of gravity. Hence the new theory is able to relate the electric and gravitational forces to one another.

This must surely represent a considerable advance. Both these long-range forces had to be modelled on the buoyancy force concept, because the alternative absorption idea gave false predictions in each case.

A curious fact emerged from both these long-range forces. Elementary particles appeared to be like low-density bubbles immersed in a sea of much higher density. Mediators must in consequence have extreme densities of perhaps 10^{47} J/m^3, or even 10^{52} J/m^3 with the contra-rotating models, so that they will have room to move. Then even the sub-components of so-called "elementary" particles, like electrons or quarks, must be fluffed out forms of physical energy. They could possibly be structured like thistledown.

In the main text a Grid having computer-like properties was deduced to exist which must pervade all space. The new electron models can throw more light upon its mesh size because this would have to be large enough to let these pass. The electrons themselves were shown to exist as particle sequences based on permanent net energy. Each member of the sequence is created from the residual energy of the previous member at facilities mounted on the Grid filaments. The place at which the replacement particle is generated does not coincide with that of the previous member, however. This is computed by the Grid using simulated wave interference patters. Where constructive interference occurs the new particle is limited to appear.

During its brief life, curiously enough, the sub-atomic particle obeys extended Newtonian mechanics just like objects of macroscopic size. It is only the sub-atomic particles as a whole, being themselves complex little systems, which jump about by repeated reconstruction. Whilst they live all particles obey the same laws of motion so that effects at the sub-atomic scale translate to those observed on the large scale. Established quantum physics is also embarrassed by a conflict with relativistic effects. Speeds close to that of light cannot be handled properly at the sub-atomic scale.   Again the new treatment eliminates the conflict and shows all effects can be handled by the same basic equations.   A unified system of mechanics applies throughout nature.

No longer is it therefore reasonable to think of quantum effects as being in conflict with classical physics.   The whole acts in unison in perfect harmony and devoid of all contradiction.

All the models described are absolutely dependent on negative energy states to complement the positive kind. Further supporting evidence for its existence has therefore been provided.

<u>REFERENCES</u>. NON-TECHNICAL SECTION

101) Abbott, F.
Baby Universes and Making the Cosmological Constant
Zero.
Nature, vol.336, 22/29 Dec. 1988, p 711.

102) Bohm, David
Wholeness and the Implicate Order.
Routledge & Kegan Paul, London, Boston & Henley,

103) Capra F.
The Tao of Physics.
Wildwood House, London.

104) Davies, Paul C. W.
The Accidental Universe,
Cambridge University Press, 1982.

105) Davies, Paul C. W.
God and the New Physics.
Penguin Books.

106) Davies, Paul C. W.
Superforce.
Heinemann, London, 1984.

107) Davies, P. C. W. & Brown, J. R. (Editors)
The Ghost in the Atom.
Cambridge University Press, 1986.

108) Davies, P. C. W. (Editor)
The New Physics.
Cambridge University Press, 1989.

109) Delgardo, P. & Andrews, C.
Circular Evidence.
Bloomsbury Publishing Ltd,
2 Soho Square, London W1V 5DE, 1989

110) Einstein, A.
Relativity, The Special and General Theory.
Bonanza Books, New York, 1952.

111) Einstein, Albert & Infeld, Leopold
The Evolution of Physics.
Cambridge University Press, 1938.

112) Epstein, Lewis Carroll
     Relativity Visualised.
     Insight Press,
     614 Vermont St. San Francisco C.A. 94107.

113) Feynman, Richard, P.
     QED, The Strange Theory of Light and Matter.
     Princeton University Press, 1985.

114) Gribbon, J.
     The Paradox of Schrödinger's Waves.
     New Scientist, 27 Aug. 1987.

115) Hawking, Stephen. W.
     A Brief History of Time.
     Bantam Press, London.

116) Josephson, Brian D., FRS, (Prof. Phys., Camb. Univ.)
     Physics and Spirituality: the Next Grand Unification?
     Phys. Educ 22(1987), IOP Publishing Ltd.

117) LeShan, Lawrence
     The Medium, the Mystic and the Physicist
     (Towards a General Theory of the Paranormal).
     Turnstone Books, London.

118) Meadows, D. H. & D. L., Randers,J. & Behrens III, W.W.
     The Limits to Growth.
     Universe Books & The Club of Rome, Leo Thorpe Ltd.,
     Wembley, 1972.

119) Mesarovic, Mihajlo & Pestel, Eduard
     Mankind at the Turning Point.
     Hutchinson of London (Club of Rome), 1974

120) Shamos, Morris H.
     Great Experiments in Physics.
     Holt Rinehart and Winston N.Y., Jan. 1960.

121) Sheldrake, Rupert
     The Presence of the Past.
     Morphic Resonance and the Habits of Nature.
     Collins, 8 Grafton St., London W1, 1988.

122) Teillard, Pierre
     The Phenomenon of Man.
     Collins.

123) Tryon, Edward P.
What Made the World?
New Scientist, 8 March 1984, pp 14-16.

124) Will, C. M.
Was Einstein Right?
Oxford University Press, 1988.

125) Wilson, Colin.
Poltergeist!
New English Library, 1981.

TECHNICAL REFERENCES (EXCLUDING THOSE OF NEGATIVE MASS)

201) Aspect, Alain, Dalibard, Jean & Roger, Gerard.
Experimental Tests of Bell's Inequalities used in
Time-varying Analysers.
Physical Review Letters, Vol.49, No.25 20 Dec.1982.

202) Blanchard, C.H., Burnett,C.R., Stoner,R.G. & Weber,R.L.
Introduction to Modern Physics.
Sir Isaac Pitman & Sons Ltd., London.

203) Chiu, Hong-Yee & Hoffmann, William, F. (Editors)
Gravitation and Relativity.
W. A. Benjamin, Inc., New York, Amsterdam, 1984.

204) Clemence, G. M.
Rev. Mod. Phys. 19, 361(1947)

205) Davies, Paul C. W.
Quantum Mechanics.
Routledge & Kegan Paul, 1984

206) Einstein, A.
Ober die Spezielle und die Allgemeine
Relativitatstheorie.
Vieweg and Sohn, Braunschweig, 1917.

207) Gibbons, G. W., Hawking, S.W. & Siklos, S.T.C.
(Editors)
The Very Early Universe.
Cambridge University Press, (Nuffield Workshop
Cambridge),   21 June 1982.

208) Heisenberg, Werner
The Physical Principles of the Quantum Theory.
Dover Publications, 1930   (Reprinted 1949).

209) Horlock, J. H. and Winterbone, D. E.
The Thermodynamics and Gas Dynamics of
Internal-Combustion Engines Vol II
Clarendon Press. Oxford.

210) Møller, C.
The Theory of Relativity. (2nd. Edition 1972).
Oxford University Press, 1972.

211) Novikov, I. D.
Evolution of the Universe.
Cambridge University Press, 1983.

212) Pearson, R. D.
Quantum Gravitation.
Ready for printing, 1989

213) Rae, Alastair I. M.
Quantum Mechanics.
IOP Publishing Ltd., Adam Hilger, Bristol, 1986

214) Shapiro, Irwin I.
The Fourth Test of General relativity.
Phys. Rev. Lett., vol. 13, No. 26, 28 Dec. 1964, p.789.

215) Starobinskii, A. A. & Ya. B. Zel'dovich.
Quantum Effects in Cosmology.
Nature, Vol. 331, 25 Feb. 1988.

216) Vallentine, H. R.
Applied Hydrodynamics.
Butterworths, London, 1967.

217) Woehler, K. E., Professor in Department of Physics,
Naval Postgraduate School, Monterey, California.
(Private Communication)

LITERATURE PERTAINING TO NEGATIVE MASS.

301) Bartlett, D. F. and Buren, Dave Van.
Equivalence of Active and Passive Gravitational
Mass Using the Moon.
Physical Review Letters, Vol. 57, 1986, pp. 21-24

302) Bondi, H.
Negative Mass in General Relativity.
Rev. Modern Physics, Vol. 29, 1957, pp. 423-428, 1957.

303) Bonnor and Swaminarayan, N. S.
An Exact Solution for Uniformly Accelerated Particles
in General Relativity.
Zeitschrift fur Physik, Vol. 177, 1964, pp. 240-256.

304) Carter, Brandon.
Global Structure of the Kerr Family of
Gravitational Fields.
Physical Review, Vol. 174, 1968, pp. 1559-1571.

305) Carter, Brandon.
Complete Analytical Extension of the Symmetry axis
of Kerr's Solution of Einstein's Equations.
Physical Review, Vol. 141, 1966, pp. 1242-1247.

306) Ciufolini, Ignazio and Ruffini, Remo
Equilibrium Configurations of Neutron Stars and the
Parametrized Post-Newtonian Metric Theories
of Gravitation.
Astrophysical Journal, Vol. 275, 1983, pp.867-877.

307) Forward, R. L.
Far Out Physics.
Analogue Science Fiction/Science Fact, Vol. 95, No. 8,
Aug 1975, pp. 147-166.

308) Forward, Dr. Robert L.
Negative Matter Propulsion.
Forward Unlimited, P.O. Box 2783, Malibu,
California 90265, USA, 30 May 1988

309) Forward, R. L.
Future Magic.
Avon, New York, 1988.

310) Kerr, R. P.
     Gravitational Field of a Spinning Mass as an Example of
     Algebraically Special Metrics.
     Physical Review Letters, Vol. 11, 1963, pp. 237-238.

311) Martins, R. de A.
     Causal Paradoxes Implied by the Hypothetical
     Co-existence of Positive and Negative Mass Matter.
     Lett.Nuovo Cimento, Vol. 28, Ser, 2, 1980, pp. 265-268.

312) Miller, B. D.
     Negative Mass Lagging Cores of the Big Bang.
     Astrophysical Journal, Vol. 208, 1976, pp. 275-285.

313) Nieto, M. M. and Bonner, B. E.
     Looking for new Gravitational Forces with Antiprotons.
     Proc. Antiproton Science and Technology Workshop,
     Santa Monica, California, 6-9 Oct 1987.
     World Scientific Publishing, Singapore, 1988.

314) Weber, J.
     General Relativity and Gravitational Waves.
     Interscience Publishers, Inc., New York, 1981,pp. 5-6.

APOLOGIES

A private communication from Denis Midgley was received just
before going to press.   He mentioned that Dr. Cullwick
published a paper which partially anticipated the derivation
for equation[2.4] given in Chapter T.S.2.   He showed that
starting with "$E = m.c^2$" it was possible to derive Einstein's
relation between relative mass and rest mass by considering
acceleration in a similar manner to that used here.   It did
not start by showing by logic that mass and energy are
equivalent, however. In the derivation given here "$E = m.c^2$"
was not assumed but appeared as part of the derivation.

It may be that other aspects of the theory have also
been partially or even fully anticipated without the
author's knowledge. If so then apologies are offered for
not giving acknowledgement.